# MODERN RECEIVER FRONT-ENDS

# MODERN RECEIVER FRONT-ENDS

## Systems, Circuits, and Integration

**JOY LASKAR**
Georgia Institute of Technology

**BABAK MATINPOUR**
VT Silicon Inc.

**SUDIPTO CHAKRABORTY**
Georgia Institute of Technology

A JOHN WILEY & SONS, INC., PUBLICATION

Published by John Wiley & Sons, Inc., Hoboken, New Jersey.
Published simultaneously in Canada.



For general information on our other products and services please contact our Customer Care Department within the U.S. at 877-762-2974, outside the U.S. at 317-572-3993 or fax 317-572-4002.

Wiley also publishes its books in a variety of electronic formats. Some content that appears in print, however, may not be available in electronic format.

***Library of Congress Cataloging-in-Publication Data is available.***

ISBN 0-471-22591-6

Printed in the United States of America.

10 9 8 7 6 5 4 3 2 1

# CONTENTS

# PREFACE

In recent years, the research and developments in the area of RF and microwave technologies have progressed significantly due to the growing demand for applicability in wireless communication technologies. Starting from 1992, wireless communication technologies have become quite mature. In the modern era of electronic developments, design of wireless handsets is an example of integration of many diverse skill sets. Classical books in the areas of microwave technology provide us with an in-depth knowledge of electromagnetic fundamentals. On the other hand, books covering analog circuit design introduce the reader to the fundamentals of basic building blocks for wireless communications. However, with the tremendous advances in wireless technologies, both in wireless systems as well as semiconductor processes, wireless solutions have become a manifestation of integrated design philosophies in the areas of analog, microwave, and communication system theory. The main focus of this book is the integration of and interaction among various building blocks and the development and characterization of receiver subsystems for wireless applications. During the years of our involvement with the graduate curriculum at the Georgia Institute of Technology, we felt that such a book would be very helpful for understanding of receiver front-end development and architectural trade-offs, and could be helpful to students, professionals, and the interested scientific community. It could also serve the needs of an aggressive researcher in the area of receiver front-ends.

The book is organized into nine chapters, as outlined below.

Chapter 1 provides an introduction to the advanced receiver architectures. Chapter 2 provides the different issues and parameters concerned with system design for receiver front-ends. Chapter 3 is targeted toward an overview of different receiver architectures and their applications. Chapter 4 walks the reader through an example of high-frequency receiver front-end design in a commercially available silicon germanium technology. Receiver design and developments are discussed in detail, along with simulation and characterization techniques necessary to focus on details of implementation to the reader, and the chapter is anticipated to be helpful for students in their research projects. This chapter provides the reader with the full flow of a design cycle, starting from computer simulation and ending with real silicon implementations. Chapter 5 is focused on various subharmonic mixing techniques, starting with the different methods of realizing subharmonic mixing technique, and followed by their integrability with other RF building blocks. Chapter 6 provides a silicon-based active subharmonic mixing approach, which is very helpful in developing an in-depth understanding for the receiver front-ends. Chapter 7 demostrates various types of passive components and their integration methodologies in silicon-based substrates. Chapter 8 describes various practical issues in the integration of receiver subsystems. Chapter 9 introduces the interested reader to the potential future applications of wireless communications.

All in all, we have tried to maintain a good balance of theoretical foundation, design procedures, and practical characterization issues toward the development of fully integrated receiver front-ends. All of the architectures are demonstrated in commercially available semiconductor process technologies. Silicon-based processes have been focused on in greater detail due to their popularity and potential in future generations of lower-cost wireless communication solutions.

This book is written with the assumption that the reader has knowledge of basic electronic circuits, microwave fundamentals, and communication theory. Although the book starts with basic receiver design and integration techniques, and talks about the state-of-the-art implementations afterward, a prior background in all of the above-mentioned areas provides a much better appreciation for the technical material presented in this book.

JOY LASKAR
BABAK MATINPOUR
SUDIPTO CHAKRABORTY

*Atlanta, GA*
*June 2003*

# ACKNOWLEDGMENTS

The authors would like to acknowledge their colleagues who have graduated from the Microwave Applications Group at the Georgia Institute of Technology including: Dr. Nicole Evers of GE R&D, Prof. Anh-Vu Pham of University of California–Davis, Dr. Sangwoo Han of Anadigics, Dr. Ramana Murty of IBM, Dr. Seungyup Yoo of RF-Solutions, Prof. Deukhyoun Heo of Washington State University at Pullman, Dr. Kyutae Lim, Dr. Stephane Pinel, Dr. Chang-Ho Lee, Dr. Sebastien Nuttinck of the Georgia Institute of Technology, Dr. Albert Sutono of Infinera, Dr. Daniela Staiculescu, Dr. Chang-Ho Lee of the Georgia Institute of Technology, Prof. Emery Chen of National Taiwan University, Dr. Hongwei Liang of Texas Instruments, Dr. Arvind Raghavan and Dr. Edward Gebara of Quellan, and Dr. Joshua Bergman of Rockwell Science Center.

This work would not have been possible if not for the creative freedom provided at the Georgia Institute of Technology under the direction of Prof. Roger Webb. In addition, the authors would like to acknowledge the support of the NSF Packaging Research Center under the direction of Prof. Rao Tummala, the Georgia Institute of Technology Microelectronics Research Center under the direction of Prof. Jim Meindl and the leadership provided by Herb Lehman, Director of Georgia's Yamacraw program.

Most importantly, the authors thank their families for their patience and support, especially Devi Laskar, Anjini Laskar, Ellora Laskar, Devrani Laskar, Soraya Matinpour, Ali Matinpour, Mala Chakraborty and Sima Chakraborty.

# 1

# INTRODUCTION

Any communication system, in the simplest form, consists of a transmitter, a signal path, and a receiver. The performance of such systems depends heavily on each of the building blocks and the impact of the given communication link on the signal. Although the impact of the path is fixed by the frequency of the RF signal and the properties of the physical medium in which the signal propagates, the behavior of the transmitter and receiver can be flexible. The electrical performances of the transmitter and receiver determine the impact of these blocks on the signal and limit the quality and range of the communication link. The appropriate topology, semiconductor technologies, and a careful design based on well-defined system parameters can make a huge difference in performance, cost, and marketability of the individual transmitter, receiver, and the entire system.

This book will take a very narrow focus on receiver design by limiting its scope to receivers for wireless applications. In order to provide the reader with a comprehensive understanding of the subject at hand, a thorough system and architecture analysis will be presented. This top level analysis will then be complemented with a touch of reality when we start describing choices, compromises, and challenges that a design engineer will face during the process of developing a cost-effective and marketable receiver product in the form of an integrated circuit (IC).

The basic developments in the area of wireless communication date back to the early 20th century. Since the early years of the 20th century, wireless engineering has come a long way. Most of the basic principles of sophisticated radio architecture, as we see it today, were developed using vacuum tubes around 1930. Starting with the basic foundation laid down by Maxwell (1883), and with subsequent inventions in wave propagation and wireless telegraphy by Hertz, Marconi, and others, wireless technology was born around 1900 in a very primitive form. Demonstration of a superheterodyne receiver by Armstrong dates back to as early as 1924.

*Modern Receiver Front-Ends.* By J. Laskar, B. Matinpour, and S. Chakraborty
ISBN 0-471-22591-6 

Armstrong's superheterodyne receiver underwent considerable refinement during the 1920s and 1930s. This was the time when radio pioneers considered the use of homodyne architectures for single vacuum tube receivers. For over two decades, the standard low-end consumer AM tunable radio used a system of five vacuum tubes [1, 2]. A major milestone was set by the invention of the transistor by Bardeen, Brattain, and Schockley in 1948, which changed the world of vacuum tubes. However, implementing radios was a farsighted vision at that time. As semiconductor technologies became more mature, more circuit integration took place. Starting with small-scale integration in the standard integrated circuits, the trend moved toward more integration and high-speed microprocessors. With the tremendous growth in the VLSI side, demands for ubiquitous computing and wireless applications increased. Vance at ITT was the first to apply direct conversion for pager applications by means of a single-chip receiver [3]. During the early 1980s, direct conversion receivers were developed at Motorola (patents were filed in the 1985 time frame) as a possible way to implement compact radios. However, the first attempt did not make a significant impact in the marketplace. To date, the superheterodyne architecture has been the winner for the industry in wireless communication technologies. Most of the direct conversion architecture realizations still reside in the research domain.

The initial necessity of communicating through short messages eventually evolved to the need to communicate audio- and video-based messages, and many other real-time applications. All of these basic requirements led to the need for increasingly higher data rates for next-generation wireless communication applications. Two primary directions driving the application space as of today include cellular telephony and wireless local area networks (WLANs). In recent years, there has been a lot of interest in more integration on-chip to realize these solutions in a compact and lower-cost fashion. Very low IF (VLIF) and direct conversion architectures are quite attractive for ultracompact, low-power, and low-cost solutions for wireless applications. If implemented successfully, direct conversion radios are the most compact realizations one can achieve. With the growing demand for wireless technology, and simultaneous development of mature and reliable semiconductor technologies, direct conversion architecture is favorable for future wireless communication technologies. At the time of writing this book, a major part of the IC industry is focused toward communication applications, both wireless and wired. Emerging applications include wideband code division multiplexing (WCDMA), IEEE802.11X, multiband, ultrawide band (UWB), and 60 GHz WLAN technology. All of these implementations are being targeted to low supply voltages as well, as a consequence of shrinking dimensions of the transistors, and to realize low-power solutions. All of these developments have motivated researchers to investigate superior process technology, novel circuit design techniques, and improved system engineering. The direct conversion architecture has the potential to satisfy the needs of most of the above mentioned applications.

Along with the development of wireless communication technologies, semiconductor technologies have also experienced tremendous evolution over the past decade. Starting with gallium arsenide (GaAs) based technologies for high-frequency design, the focus has slowly shifted from III–V semiconductors to silicon-based technologies for lower cost and higher integration during the early 1990s in the research domain. Currently available silicon-based technologies, which show enormous potential for RF technologies, include standard digital CMOS, silicon-on-insulator (SOI), and silicon–germanium (SiGe). Bulk CMOS technologies have become much more attractive for RF design during recent times because of low cost and other potential advantages related to continued scaling in the deep-submicron (DSM) regime. RF circuit implementations in standard CMOS technologies have developed considerably well over the past couple of years [4]. However, there still exists a question about how much of the RF CMOS implementation will be really adopted by the industry. This skepticism is the result of the lower yield and reliability of such technologies for high-frequency analog and RF applications. SOI-based CMOS processes have shown improvement over standard digital CMOS technologies in many aspects [5]. Silicon-on-insulator technology has shown promise in digital microprocessor applications and, hence, is a strong candidate for future system-on-a-chip (SOC) realizations. Although SOI technology was originally proposed in 1970, its potential has not yet been fully explored in analog/RF designs. SiGe BiCMOS technology has proven to be a very strong candidate for RF as well as back-end digital designs [6]. Recent reports of state-of-the-art SiGe BiCMOS technologies [7] have shown a cutoff frequency ($F_t$) of up to 200 GHz and are quite capable of handling next-generation high-data-rate applications. Currently, silicon-based RFIC solutions are just about to penetrate the commercial market for both wired and wireless applications. At this point, it would be quite interesting to take a step back and think carefully about advances in the wireless IC world. It is noteworthy that the very basic principles of circuit design have not changed significantly from the 1920s, but their applications have. Experts are often tempted to call this technical advancement an "evolution" as opposed to a "revolution."

## 1.1. CURRENT STATE OF THE ART

Starting with the initial developments of superheterodyne architectures, radio front-ends have gone through many changes. Many researchers are working mostly toward implementing very low IF frequency and direct conversion radio front-ends for low power and compact implementations. Radio frequency receiver integrated circuits require a combination of expertise in the areas of circuit design and system architecture, and the choice of a suitable process technology for various applications. As we progress through the subsequent chapters, the reader will get a clear picture about how these disciplines are intertwined in today's world of advanced receivers. Table 1.1 summarizes reports of direct conversion solutions to date, along with key distinguishing technological features. It is quite interesting to note the co-existence of the III–V semiconductor (GaAs MESFET) implementations with those in silicon-based technologies.

This book will address the different issues and challenges in the development of highly integrated receiver subsystems. It reviews a wide variety of receiver subsys-

**Table-1.1.** Summary of reported state-of-the-art direct conversion solutions:

| References | Technical Approach | Process Technology | Application |
|---|---|---|---|
| Reference [10, 17] (UCLA) | 1. Active circuitry in the front-end<br>2. Single balanced mixer | 0.6 μm CMOS | GSM (0.9 GHz), 1.9 GHz |
| Reference [11] (Toshiba R&D) | 1. Active circuitry in the front-end | 0.8 μm SiGe BiCMOS | PCS, 1.9 GHz |
| Reference [8] (Mitsubishi) | 1. Passive-circuit-based front-end<br>2. Subharmonic APDP-based topology | GaAs MESFET, SiGe | WCDMA |
| Reference [12] (Helsinki U. of Technology) | 1. Active circuitry in the front-end<br>2. Uses off-chip inductors<br>3. Modified Gilbert cell mixer | 0.35 μm SiGe BiCMOS | WCDMA |
| Reference [9] (Georgia Tech) | 1. Subharmonic APDP-based topology | GaAs MESFET | C band, 5.8 GHz |

tem designs, both in terms of topology as well as semiconductor technology. It covers a detailed design perspective, from systems to circuits at high frequency, with a focus on implementation in low-cost semiconductor technologies. It also provides the practical challenges faced by the designers in carrying out a fully integrated receiver solution, with a look forward to futuristic applications in the areas of wireless communications.

## REFERENCES


1. T. H. Lee, *The Design of CMOS Radio-Frequency Integrated Circuits,* Cambridge University Press, 1998.
2. W. R. Maclaurin, *Invention and Innovation in the Radio Industry,* Macmillan, NY, 1949
3. I. A. W. Vance, "Fully integrated radio paging receiver," *IEEE Proc., 129,* part F, 1, 2–6, 1982.
4. P. H. Woerlee, M. J. Knitel, R. van Langevelde, D. B. M. Klaassen, L. F. Tiemeijer, A. J. Scholten, and A. T. A. Zegers-van Duijnhoven, "RF CMOS performance trends," *IEEE Transactions on Electron Devices, 48,* 8, 1776–1782, August 2001.
5. C. L. Chen, S. J. Spector, R. M. Blumgold, R. A. Neidhard, W. T. Beard, D. -R. Yost, J. M. Knecht, C. K. Chen, M. Fritze, C. L. Cerny, J. A. Cook, P. W. Wyatt, and C. L Keast, "High-performance fully-depleted SOI RF CMOS," *IEEE Electron Device Letters , 23,* 1, 52–54, Jan. 2002.

6. J. D. Cressler, "SiGe HBT technology: A new contender for Si based RF and microwave applications," *IEEE Transactions on Microwave Theory and Techniques, 46,* 5, part 2, 572–589, May 1998.
7. *http://www. ibm. com/news/2001/06/25. phtml*
8. M. Shimozawa, T. Katsura, N. Suematsu, K. Itoh, Y. Isota, and O. Ishida, "A Passive-type even harmonic quadrature mixer using simple filter configuration for direct conversion receiver," in *IEEE International Microwave Symposium,* pp. 517–520, June 2000.
9. B. Matinpour, S. Chakraborty, and J. Laskar, "Novel DC-offset cancellation techniques for even-harmonic direct conversion receivers," *IEEE Trans. Microwave Theory and Tech., 48,* 12, 2554–2559, December 2000.
10. B. Razavi, "A 900 MHz CMOS direct conversion receiver," in *Symposium of VLSI Circuits, Digest of Technical Papers,* pp. 113–114, 1997.
11. S. Otaka, T. Yamaji, R. Fujimoto, and H. Tanimoto, "A very low offset 1. 9 GHz Si mixer for direct conversion receivers," in *Symposium of VLSI Circuits, Digest of Technical Papers,* pp. 89–90, 1997.
12. J. Jussila, J. Ryynanen, K. Kivekas, L Sumanen, A. Parssinen, and K. A. I. Halonen, "A 22-mA 3. 0-dB NF direct conversion receiver for 3G WCDMA," *IEEE Journal of Solid-State Circuits, 36,* 12, 2025–2029, December 2001.
13. C. D. Hull, J. L. Tham, R. R. Chu, "A direct-conversion receiver for 900 MHz (ISM band) spread-spectrum digital cordless telephone," IEEE Journal of Solid-State Circuits, vol. 31, no. 12, pp. 1955–1963, Dec. 1996.

# 2

# RECEIVER SYSTEM DESIGN

This chapter provides a detailed illustration of system-level parameters for receiver front-end design. A thorough system analysis is the first step in designing a receiver subsystem. The system design helps to define the specification and scope of the entire receiver and every building block in the lineup. This will in turn determine which components will be integrated on-chip or implemented off-chip by other means. Issues such as frequency scheme and the interface with other components in the overall communication system are some of the top-level items determined in this phase. These will allow us to define an overall specification for the receiver, after which the analysis can be extended to the individual blocks of the receiver.

## 2.1. FREQUENCY PLANNING

Design of a frequency plan has a direct dependence on the receiver topology, number of down-conversions, and the modes of operation, which can be simplified to simplex or duplex operation. This section will focus on the common superheterodyne topology operating in a duplex system. The block diagram of such receiver is shown in Fig. 2.1. There are only two down-conversions, one from RF to IF and the other from IF to baseband I and Q. A thorough frequency planning will involve study of blockers, spurs, and image frequencies. Blockers, spurs, and image interferers are RF signals that are transmitted into the air by other wireless devices and can penetrate and either saturate the receiver or interfere with the signal being received.

*Modern Receiver Front-Ends.* By J. Laskar, B. Matinpour, and S. Chakraborty
ISBN 0-471-22591-6 © 2004 John Wiley & Sons, Inc.

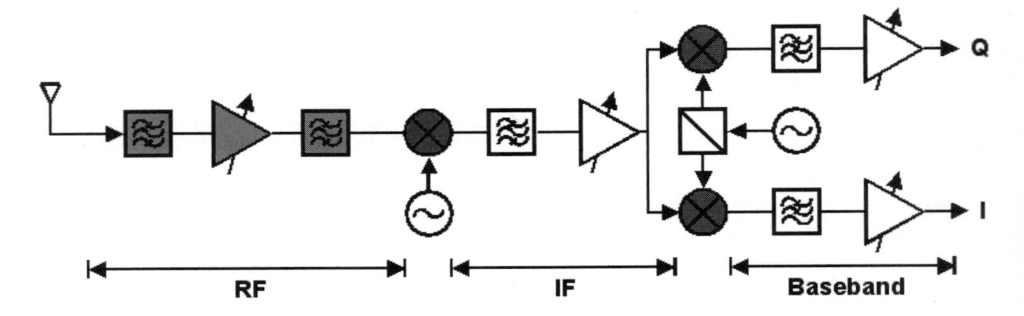


**Fig. 2.1.** Block diagram of a common superheterodyne receiver.

### 2.1.1. Blockers

Understanding the wireless applications that coexist in the frequency spectrum surrounding the band of interest is one of the very important steps in determining a sound frequency plan. Applications that use high-power transmitters can create problems by saturating the receiver front-end and potentially damaging such components. Applications such as mobile phone services that are widely deployed and operate with handset output powers in excess of 1 W are especially troublesome. Operating close to such frequency bands places great demands on front-end filter selectivity. Designers must also be careful in using such frequencies as an IF for a higher-frequency application. Many satellite receivers use L-band frequencies for IF but avoid using the bands occupied by the mobile phones as an IF frequency to avoid interference.

Fig. 2.2 shows some of the existing commercial applications below 6 GHz. In addition to the applications shown here, there are also additional government and military bands, especially at 10 GHz, that operate high-power radars and have to be considered. Although some of these higher-frequency blockers experience higher air propagation loss, they typically operate at higher output power. Radar applications can easily exceed a few watts of output power.

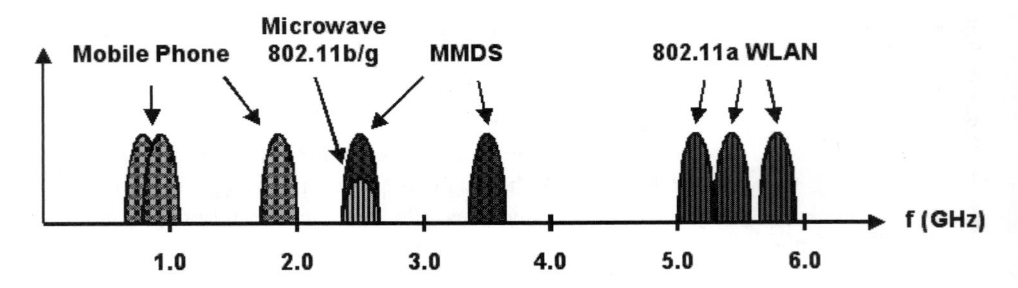


**Fig. 2.2.** Frequency spectrum showing various applications and potential blockers.

Fig. 2.3 shows the propagation of a blocker signal through the front-end of a receiver for three different cases depending on the relative frequency of the blocker to the receive band. In case (a), the frequency of the blocker is very close to receiver band of interest and it experiences little filtering in the band-pass filter. This places stringent linearity requirements on the LNA and RF mixer, which has a consequent impact on power consumption of these components. In case (b), the blocker frequency is significantly lower than the receive band and it experiences adequate rejection through the filters. In case (c), the blocker frequency is the same distance from the receive band as case (b), but this time it is located at the higher side of the band. Assuming the same filter rejection for this higher-frequency blocker, the blocker level is further attenuated by the high-frequency gain roll-off in the active components in the front-end, such as the LNA and RF mixer.

The designer needs to perform this analysis for every potential blocker in the spectrum and make adequate corrections either to the frequency plan or to the filter specification. Unfortunately, the frequency plan does not provide many options for alleviating the impact of blockers on the receiver RF front-end. The location of the receive band is typically predetermined by standards and cannot easily be shifted, leaving the filtering as the only means for addressing the blockers in the RF front-end. This is not the case for the IF frequency. In most cases, the designer can chose the IF frequency such that it avoids blockers that can interfere with the IF chain.

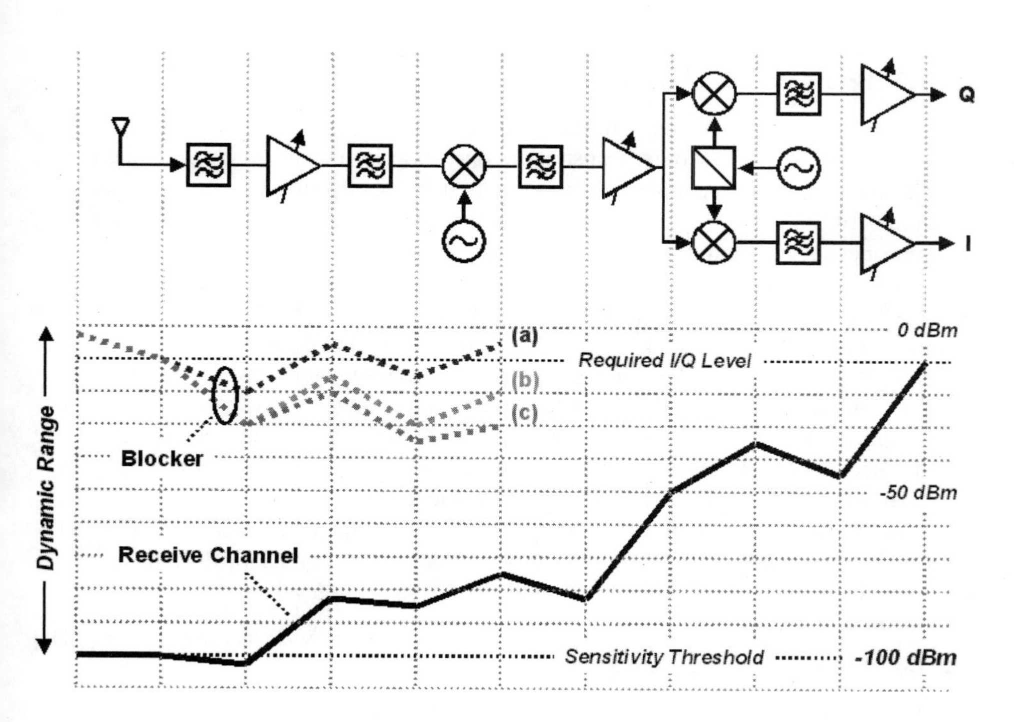


**Fig. 2.3.** Block diagram showing the propagation of a blocker through the chain.

The only limitation for selection of an IF frequency would then become the availability of IF surface-acoustic wave (SAW) filters at the chosen frequency, if such filters are indeed required.

The formula below describes the rejection required for the front-end band-pass filters (BPFs). A margin (MRG) is typically needed to determine the blocker back-off from the input 1-dB compression point ($IP_{1dB}$) of the receiver and its individual components. The input blocker level (IBL) is the blocker strength at the antenna output. This number has already been adjusted for path loss and antenna selectivity.

$$BPF_{rej} = (IBL + MRG - IP_{1dB_{LNA@f(blocker)}}) + (IP_{1dB_{LNA}} + G_{LNA@f(blocker)} - IP_{1dB_{Mixer@f(blocker)}}) \quad (2.1)$$

***Example from Fig. 2.3(c):***

$IBL = -10$ dBm
$MRG = 10$ dB
$IP_{1dB_{LNA@f(blocker)}} = -20$ dBm
$IP_{1dB_{Mixer@f(blocker)}} = -20$ dBm
$G_{LNA@f(blocker)} = 25$ dB

$$BPF_{rej} = (0 + 10 + 20) + (-20 + 25 + 20) = 45 \text{ dB}$$

### 2.1.2. Spurs and Desensing

Spur analysis is an extension of the analysis performed for blockers. Here, we do not just study the impact of other transmitters operating in the surrounding frequency bands but also study the unwanted spurious frequencies that are generated by interaction between various components of our own transceiver. This includes the interactions between the low-frequency crystal oscillator used for the synthesizer, RF and IF local oscillators, and transmit and receive signals. This analysis is performed to identify spurs that land in either RF, IF, or LO frequency bands. The spurs that register with high power levels relative to the signal of interest in these bands can be very troublesome and have to be addressed early in the frequency plan. Typically, the spurs that interfere with the LO frequencies are less problematic since the LO signals are significantly higher in power when compared to the spurs. On the contrary, the IF and especially the RF signal, which are typically low in power, are very susceptible to spur interference. It is usually best to design the frequency plan to avoid spurs that fall directly on the RF or IF bands. Desensing is one of the outcomes of such interference where a spur with a higher power level than the desired RF or IF signal lands either directly in the band or adjacent to the band and saturates the transceiver.

### 2.1.3. Transmitter Leakage

Leakage in the transmitter is a major concern for any advanced RF subsystem, especially in duplex systems. The transmitter, operating at a high power level, requires stringent filtering to maintain confinement to the transmit band and avoid interference with the adjacent bands, especially the receiver. Transmitter leakage into the receiver can result in desensing of the receiver by saturating the receiver front-end or causing oscillations.

As shown in Fig. 2.4, the transmit signal can leak into the receiver input and cause an oscillation by finding a leakage path after the front-end gain. This is a major issue that makes it very difficult to integrate both transmit and receive functions of a duplex system on a single chip.

### 2.1.4. LO Leakage and Interference

Local oscillator signals and their harmonics are major sources of spurious interference. As shown in Fig. 2.5, there are many potential paths available for LO leakage and interference. These leakage paths are created either through the IC substrate and package or through the board on which the IC is mounted. It is often very difficult to identify and address every leakage path that may exist. Therefore, the best method for avoiding interference is to devise a frequency plan in which all LO frequencies and their harmonics, and even frequencies resulting from higher-order mixing of these signals, do not fall in the RF or IF bands. Since LO power levels are relatively higher than RF and IF levels, it is very likely to have a higher-order term involving an LO signal create significant interference or overpower the RF input signal and desense the receiver.

For the cases in which an LO1 reaches the input of the LNA, the interfering signal is amplified by the LNA, making it even a larger interferer. This typically results in saturation of the RF mixer or any other active element that follows the LNA. Since the frequency selectivity of the typical LNA is not very significant and

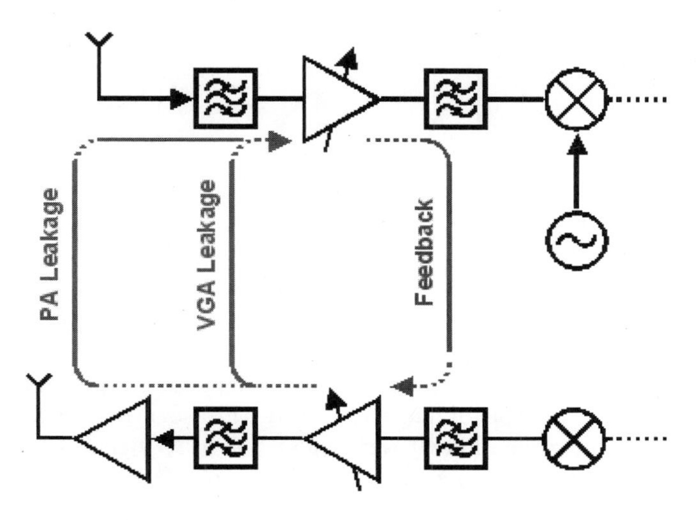


**Fig. 2.4.** Paths for transmitter leakage into the receiver and potential feedback.

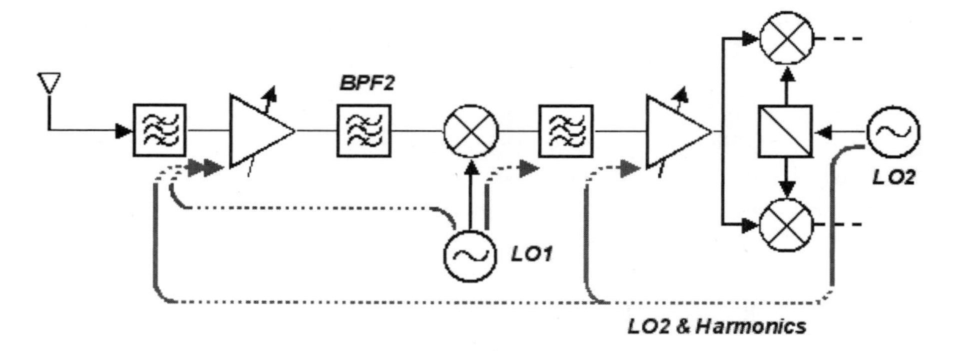


**Fig. 2.5.** Block diagram showing potential leakage paths for the LO.

there is no protection provided from the first BPF, this type of interference heavily depends on the amount of LO or LO harmonic coupling, either through the substrate or the board.

Again, the frequency plan and careful selection of IF and LO frequencies can play an important role in alleviating this problem. A good way of looking at the extent of this problem is to estimate the level of desired isolation using the inequality shown in the example below. Once the needed isolation is calculated, the designer can determine if this level of isolation is practical given the IC process, packaging, and board characteristics. Local oscillator power at the mixer input (PLO) and input 1-dB compression point ($IP_{1dB}$) of the blocks, and a safety margin (MRG) are used in the calculations:

$$\text{Isolation} > G_{\text{LNA@f(LO)}} + G_{\text{BPF2@f(LO)}} + \text{MRG} + P_{\text{LO}} - IP_{\text{1dBMixer}} \tag{2.2}$$

***Example from Fig. 2.5:***

PLO = –5 dBm

MRG = 10 dB

$IP_{\text{1dB}_{\text{Mixer@f(LO)}}} = -20$ dBm

$G_{\text{BPF2@f(LO)}} = -20$ dB

$G_{\text{LNA@f(LO)}} = 25$ dB

$$\text{Isolation} > 25 - 20 + 10 - 5 + 20 < 30 \text{ dB (very practical)}$$

Fig. 2.6 demonstrates how a second LO signal can generate harmonics that can fall in or close to the RF band of frequency. There can also be potential problems if the harmonics of the second LO interfere with the first LO signal. To avoid these two undesirable scenarios the frequencies allocated for LO1, LO2, and RF bands should be selected so that they are not integer multiples of one another. The alloca-

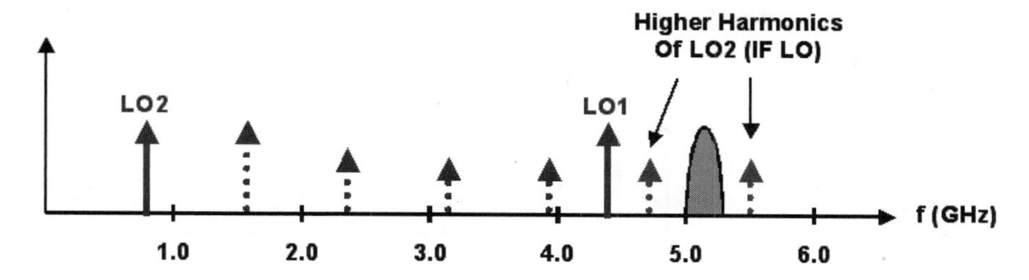


**Fig. 2.6.** Frequency spectrum showing the impact of LO harmonics.

tion of the LO frequencies should be done in such a way that their harmonics completely clear the entire RF band with a reasonable margin.

### 2.1.5. Image

Image frequency is one of the most problematic issues in designing traditional superheterodyne receivers, which is the case under study in this chapter. The image problem can be avoided by using another receiver topology with its own set of challenges.

As shown in Fig. 2.7, the image signal is located on the opposite side of the LO frequency and folds on top of the IF band as the signal is down-converted in a mixer. This creates a serious interference issue that needs to be addressed using either filtering or image-reject mixing topologies.

### 2.1.6. Half IF

Interference of half-IF frequencies is also another issue that plagues most receiver topologies. As shown in Fig. 2.8, the half-IF frequency is located directly between the LO and the RF. This half-IF signal can double in the LNA or RF amplifiers in the front-end and get down-converted into the IF band by mixing with the second

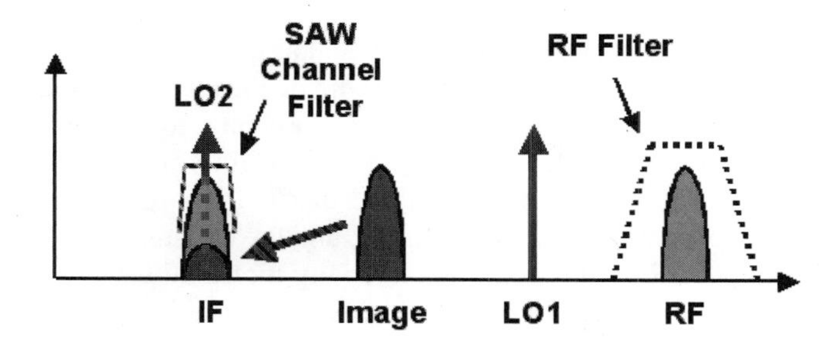


**Fig. 2.7.** Frequency spectrum showing image interference.

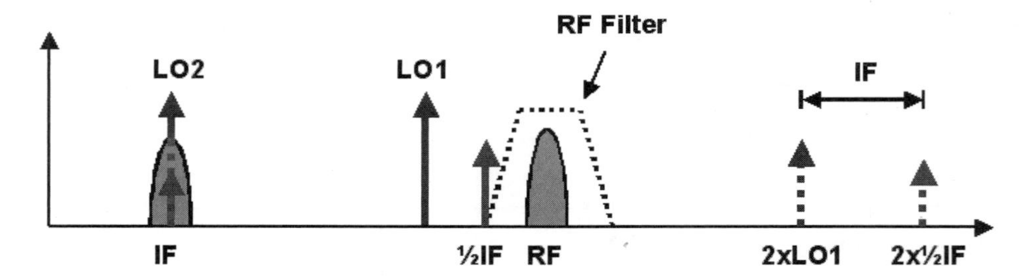


**Fig. 2.8.** Frequency spectrum showing half-IF interference.

harmonic of the LO signal. This problem can be avoided with adequate filtering in the front-end or low-distortion LNA and RF amplifier designs. By using active front-end components with low, even-order distortion products, the half-IF frequency will no longer produce a significant second harmonic, eliminating the concern for interference.

Equations 2.3 and 2.4 describe the interfering component that arises from the interaction between half-IF and LO harmonics for the case of a low-side injection mixer.

$$F_{\frac{1}{2}\text{IF Interference}} = \text{F}_{\text{RF}} - \tfrac{1}{2} \times F_{\text{IF}} \quad (2.3)$$

$$F_{\text{IF}} = (2 \times F_{\text{LO}}) \times (2 \times F_{\frac{1}{2}\text{IF Interference}}) = \text{Interference} \quad (2.4)$$

## 2.2. LINK BUDGET ANALYSIS

The purpose of a link budget analysis is to determine the individual specifications of the receiver blocks. This analysis is dependent on several key system parameters such as sensitivity, dynamic range, and input signal range required for the analog-to-digital converter (A/D) or limiting amplifier terminating the back-end of the receiver.

The basic purpose of a modern receiver is to detect and deliver an RF signal from an antenna to an A/D while maintaining signal quality as much as possible. Sensitivity and dynamic range of a receiver are the two main parameters that define the range of input RF power that must be received, and bit error rate and symbol error rate are the measures that define the acceptable quality of the received signal. Sensitivity defines the lowest input RF signal that must be detected and distinguished by the receiver with acceptable quality, and the dynamic range defines the entire range of input RF power from the sensitivity threshold up to the maximum detected signal. A link budget analysis uses these given criteria to determine the receiver lineup and the requirements of various receiver blocks. This typically involves calculations for gain, noise figure, filtering, intermodulation products (IM), and input 1dB compression ($P_{1dB}$). In this section, we will identify these components and describe the common methods used to quantify them.

### 2.2.1. Linearity

Linearity is the criterion that defines the upper limit for detectable RF input power level and sets the dynamic range of the receiver. Linearity is mostly characterized by two performance parameters: third-order intermodulation product (IM3) and second-order intermodulation product (IM2). These two parameters are the result of a two-tone analysis in which two in-band signals are subjected to the receiver or one of its components. The tones mix due to the nonlinear elements in the receiver or receiver component and generate products that can be used to characterize the extent of the nonlinearity.

Equations 2.5 and 2.6 describe the generation of IM2 and IM3, and the input intercept points for both of these types of nonlinear behavior. Whereas the intermodulation products (IM2 and IM3) are dependent on the signal power and cannot be used independently to describe performance, the intercept points are indeed independent parameters that can be used to quantify the linearity.

$$\text{IM2} = \text{A2} \times \text{RFin}^2 \tag{2.5}$$

$$\text{IM3} = \text{A3} \times \text{RFin}^3 \tag{2.6}$$

where
A2 = measure of device second-order nonlinearity
A3 = measure of device third-order nonlinearity

The relationship described above dictates a 2:1 slope for the IM2 and a 3:1 slope for the IM3 products, as shown in Fig. 2.9. This relationship can then be used in Equations 2.7 and 2.8 to determine the input intercept points for the second-order (IIP2) and third-order (IIP3) products. Output intercept points for the second-order (IP2) and third-order (IP3) products can also be calculated easily by adding the gain of the cascaded blocks to the appropriate input intercept points.

$$\text{IIP2 [dBm]} = \text{RFin [dBm]} + \Delta\text{IM3/2 [dB]} \tag{2.7}$$

$$\text{IIP3 [dBm]} = \text{RFin [dBm]} + \Delta\text{IM2 [dB]} \tag{2.8}$$

Fig. 2.9 shows the result of a two-tone power sweep with fundamental signal and its intermodulation products plotted as a function of input RF power. The intercept points are extrapolated using the plotted data. This graph can be generated by either simulation or measurement to determine a measure of linearity for a receiver or any of its components. For a link budget analysis, these numbers are then used in a lineup described by Equation 2.9 to determine the impact of individual components on the linearity of the overall receiver.

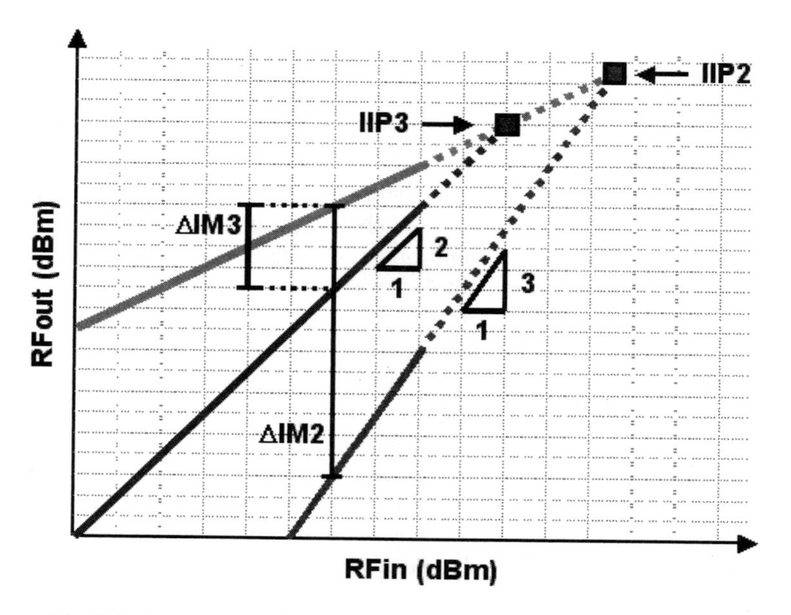


**Fig. 2.9.** Power sweep showing third- and second-order intercept points.

$$1/\mathrm{IP3}_{\mathrm{overall}} = 1/\mathrm{IP3}_1 + G_1^2/\mathrm{IP3}_2 + \ldots \tag{2.9}$$

An easy way to approach linearity is to understand the formulas but not get too attached to them. By understanding the formula above, one realizes that the overall linearity is highly dependent on the linearity of the limiting component. For example, in a receiver lineup the mixer typically becomes the limiting component. This means that by improving the linearity of the other components such as the LNA, the designer cannot see much of an overall improvement. In this example, the mixer should get most of the attention.

Gain switching is often utilized in the front-end of the receiver to ease linearity constraint and improve intermodulation performance. When the input RF signal is high, the receiver no longer needs to amplify the input signal as much; thus, it can lower the gain of the LNA or any other front-end amplifiers. This will reduce the input power into other following components, such as mixers, and avoid saturation and generation of unwanted intermodulation products.

### 2.2.2. Noise

The noise performance of the receiver defines the sensitivity of the receiver by limiting the lowest input RF power that can be detected by the receiver. There are many sources of noise that can contribute to the quality of signal in a receiver. In this section, we will discuss several major noise sources and methods of calculation. Although the noise performance of a receiver is dependent on the impact of all of the noise sources, it is very beneficial to understand when and where each noise contributor dominates the other sources and becomes the sole reason for poor signal integrity.

***2.2.2.1. Receiver Thermal Noise.*** Thermal noise is a function of random movement of particles in the medium in which the signal is traveling. The topic of thermal noise has been covered extensively in many other references, so we will only describe it briefly and highlight the relevant formulas required for a receiver system analysis. As shown in Equation 2.10, the thermal noise power is dependent on the signal bandwidth and temperature of the medium. Naturally, the noise power increases with increasing temperature and bandwidth.

$$P_n = kTB \tag{2.10}$$

where
$k$ = Boltzman's constant = $1.38 \times 10^{-23}$ [J/K]
$T$ = temperature [K]
$B$ = bandwidth [Hz]

Thermal noise of a receiver is typically referred to the input of the chain in the form of either an overall system noise temperature or noise figure. Frii's formula for calculation of receiver noise figure is shown below. This formula correlates with the lineup in Fig. 2.10.

$$F = F_{\mathrm{BPF1}} + (F_{\mathrm{LNA}} - 1)/G_{\mathrm{BPF1}} + (F_{\mathrm{BPF2}} - 1)/(G_{\mathrm{BPF1}}G_{\mathrm{LNA}}) + (F_{\mathrm{MIX}} - 1)/(G_{\mathrm{BPF1}}G_{\mathrm{LNA}}G_{\mathrm{BPF2}}) \tag{2.11}$$

where
$G$ = gain [dimensionless]
$F$ = noise figure [dimensionless]

It is very critical to note that this formula is only valid and useful when the input signal is at the threshold of sensitivity. As soon as the input signal becomes larger, the noise power of the input signal can dominate the thermal noise of the receiver and eliminate the impact of receiver thermal noise on the received signal quality. In such a case, the quality of the received signal is no longer limited by the receiver noise but by the quality of the transmitter from which it originated. This is demonstrated in Fig. 2.10. In this figure, the heavy, solid lines represent the signal level and the shaded areas describe the noise contributed by different blocks in the receiver. The light lines in the shaded areas represent the noise contributed by the LNA and the heavy lines in the shaded areas represent the noise contributed by the mixer. It is also important to note that the noise contribution of each block is divided into two components: the thermal noise and the noise contributed by the active components of the block. All passive components are assumed to be at the thermal noise level since they do not have any other internal sources of noise.

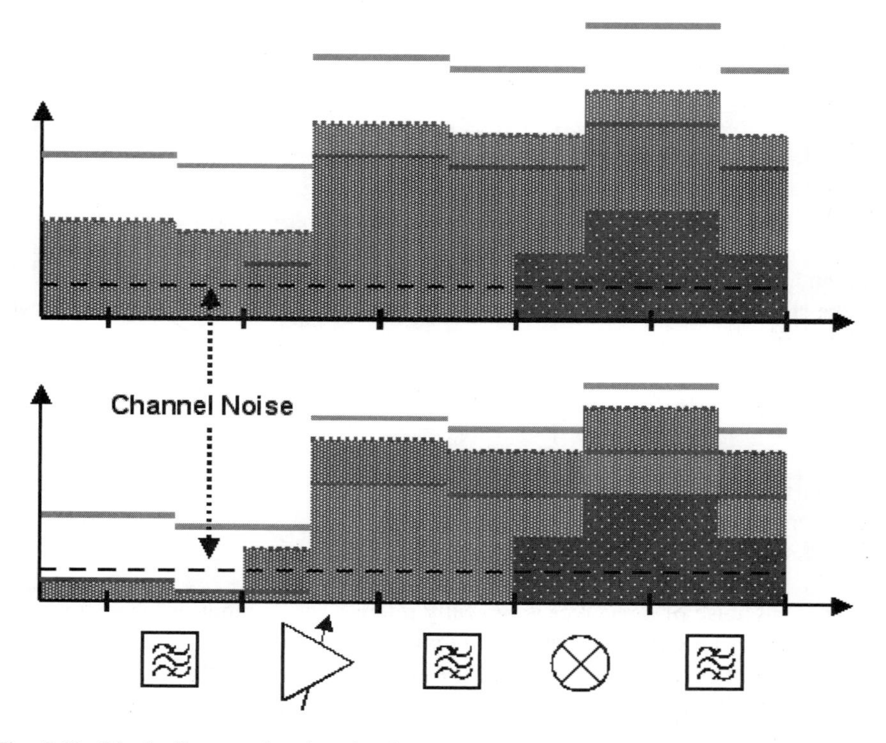


**Fig. 2.10.** Block diagram showing the decreasing impact of the noise figure as it moves down the chain.

The top diagram in Fig. 2.10 shows the case in which the noise of the input signal is significantly higher than the thermal noise level. This case typically occurs when the receiver is close to the transmitter. In such a case, it is easy to observe that the noise of the input signal dominates the thermal noise and noise contributed by the various components of the receiver; therefore, there is no degradation in the quality of the signal. The lower diagram of Fig. 2.10 describes the case in which the input signal is weak and its noise level is well below the thermal noise. In this case, the contribution of the thermal noise can significantly degrade the quality of the received signal as it propagates through the chain. However, this degradation is mostly done in the first few blocks of the receiver where the signal level is low. It is shown that once the aggregate noise that appears at the LNA and the desired signal are amplified to a higher level well above the noise of the following stages, the impact of those following stages is eliminated. This agrees with the noise figure formula described earlier.

***2.2.2.2. Transmitter Noise.*** As shown in Fig. 2.11, the broadband noise generated by the power amplifier (PA) can overcome the thermal noise of the receiver,

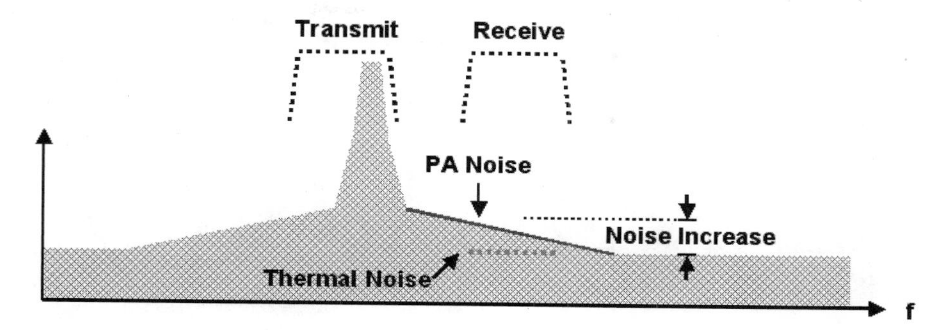


**Fig. 2.11.** Transmitter noise leaking into the receive band.

and result in a significant increase in the noise floor and, consequently, limit the sensitivity of the receiver. Apart from filtering or use of a better power amplifier, the only other choice for addressing this problem is moving the receive band further away from the transmit band.

***2.2.2.3. Phase Noise.*** Phase noise is another important noise component that should be considered for noise calculations. Phase noise, which is attributed to the local oscillator, is introduced during the mixing process. When the signal is down-converted in the mixer, the phase noise of the LO is added to the existing noise of the incoming RF signal, which is down-converted into the IF band along with the RF signal.

Phase noise is described as a relative measure of the difference between the peak LO power and the noise floor of the LO as a function of frequency offset. Phase noise contribution of the LO is calculated by integrating the LO noise power over the RF signal bandwidth. As shown in Fig. 2.12, the impact of phase noise is much more adverse closer to the LO frequency at a lower offset frequency. The phase noise typically flattens out further from the LO frequency. The close-in phase noise of the LO is typically dependent on the loop response and the phase detector performance of the phase lock loop (PLL) synthesizer, whereas the far-out phase noise is dependent on the phase noise performance of the voltage controlled oscillator (VCO). This generates a different set of requirements for different local oscillator sources. As shown in Fig. 2.12, the critical component of the LO1 source will be the VCO, whereas the critical component of the LO2 source will include the PLL phase detector and loop filter in addition to the VCO. As with all other noise sources, providing adequate gain prior to the mixer will help relax the phase noise performance of the LO sources.

**2.2.3. Signal-to-Noise Ratio.** The signal-to-noise ratio (SNR) and bit error rate (BER) are the key parameters that define the performance of the receiver. As shown in Eqns. 2.12 and 2.13, the signal-to-noise ratio is a simple measure describ-

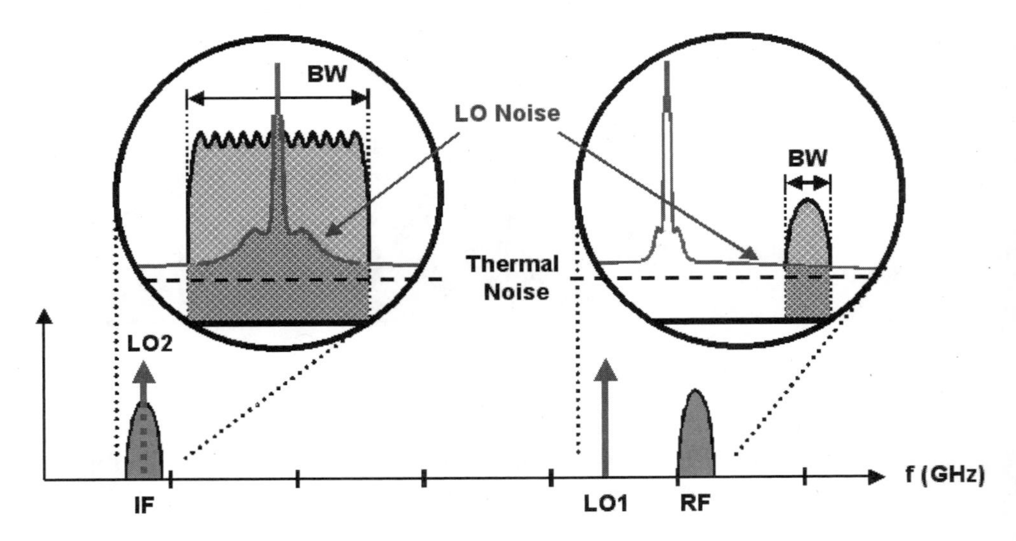


**Fig. 2.12.** Zoom on the skirt of the LO and describing the integrated noise.

ing the difference between the signal power and the noise floor. This measure is sometimes modified to include the interference and described as the signal-to-noise and interference ratio (S/N+I). Signal-to-noise ratio is used to define energy-to-noise (Eb/No) parameter needed to predict the BER performance of a receiver. The relationship between SNR and Eb/No is dependent on the modulation scheme of the received signal and described in detail in the literature [1, 2].

$$\text{SNR} = \text{signal/noise} \tag{2.12}$$

$$\text{SNR [dB]} = \text{signal [dBm]} - \text{noise [dBm]} \tag{2.13}$$

### 2.2.4. Receiver Gain

One of the most important parameters in the receiver is the overall receiver gain and the range for gain variation. As shown in Fig. 2.13, most modern receiver chains start with the antenna and end with an analog-to-digital converter (A/D). Therefore, it becomes very important for a receiver designer to not only understand the limitations of the input power and noise at the receiver input, but also consider the requirements and limitation of the A/D. The dynamic range of the A/D is the key criterion that defines the range of voltages that can be digitized with acceptable quality. The dynamic range is defined by the number of bits and the maximum input voltage swing of the A/D. The number of bits defines the lowest voltage that can be detected by the A/D and sets the maximum gain required by the receiver to boost the input RF signal from the antenna to this minimum

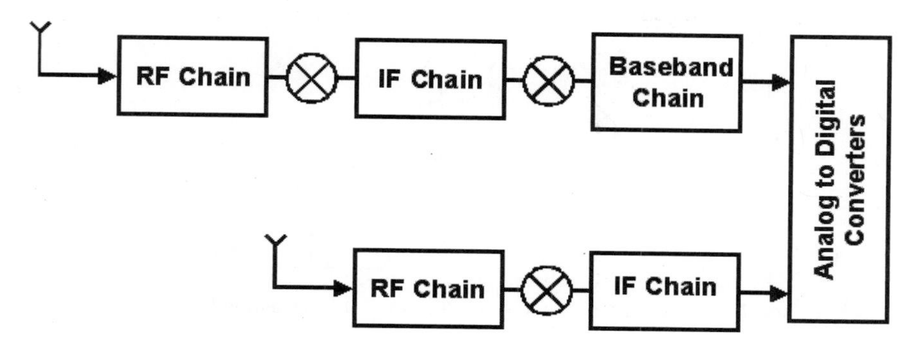


**Fig. 2.13.** Block diagram showing two example receiver lineups.

voltage level at the A/D input. The maximum input voltage swing allowed to the A/D sets the minimum gain required by the receiver. This relationship is described in the formulas below.

$$G_{\mathrm{L}} = \text{minimum receiver gain [dB]} = \mathrm{Pin}_{\mathrm{A/D,\,min}} - RS \tag{2.14}$$

$$G_{\mathrm{H}} = \text{maximum receiver gain [dB]} = \mathrm{Pin}_{\mathrm{A/D,\,max}} - RS - DR \tag{2.15}$$

where
$RS$ = receiver sensitivity [dBm]
$DR$ = receiver dynamic range [dB]
$\mathrm{Pin}_{\mathrm{A/D,\,max}}$ = maximum input signal to A/D [dBm]
$\mathrm{Pin}_{\mathrm{A/D,\,min}}$ = minimum input signal to A/D [dBm]

This gain variation range of the overall receiver needs to be spread across various components in the lineup. This gain variation is typically delegated to a switched-gain LNA or RF amplifier in the receiver front-end and one or two low-frequency variable gain amplifiers (VGA) in the IF or baseband chain. Due to ease of design and implementation, the majority of the gain variation is set for the low-frequency VGAs.

## 2.3. PROPAGATION EFFECTS

Signal propagation is an external phenomenon that does not occur in the receiver but its effects have significant impact on receiver signal integrity. In addition, better understanding of the overall system from the transmitter, through the propagation medium, and to the receiver can provide the designer with very useful insight. In our discussion, we will consider an air medium and study the impact of air and other stationary and moving objects in the path of the signal.

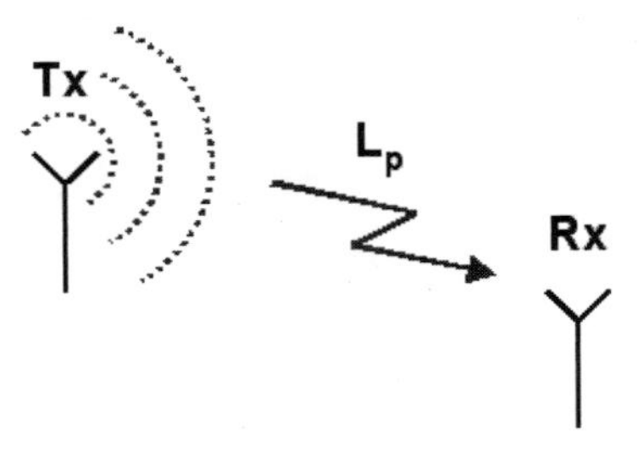


**Fig. 2.14.** A communication link showing propagation loss.

### 2.3.1. Path Loss

When a source transmits in all direction from an isotropic antenna, the signal is propagated in a spherical pattern, as shown in Fig. 2.14.

This implies that a given electromagnetic energy is spread over the surface of a sphere that grows in radius as the signal moves further away from the antenna. This causes the signal density to attenuate over distance, resulting in a signal power loss, which is referred to as path loss ($L_P$). Path loss can be calculated using the formula below.

$$L_P = 20 \log (4\pi R/\lambda) \text{ [dB]} \tag{2.16}$$

where
$R$ = path length [m]
$\lambda$ = wavelength [m]

In addition to regular path loss through the air medium, the signal can also be attenuated by rain or high water vapor concentration in the air. The water molecules absorb the electromagnetic energy, resulting in a frequency-dependent attenuation through the air. Water vapor attenuation peaks at around 2 GHz, which is the resonant frequency of the water molecules. This information has been experimentally calculated and available in the literature [1].

The relationship that describes the propagation of a signal from the transmitter, through the air, and to a receiver is described in the formula below. This relation also accounts for antenna gain in case the antenna is not isotropic.

$$P_R = P_T + G_{T,\,ANT} - L_P + G_{R,\,ANT} \text{ [dBm]} \tag{2.17}$$

where
$P_T$ = transmitter output power [dBm]
$P_R$ = receiver input power [dBm]

$\text{EIRP} = P_T + G_{T,\,ANT}$ = radiated transmitter power [dBm]
$G_{R,\,ANT}$ = receive antenna gain [dB]
$G_{T,\,ANT}$ = transmit antenna gain [dB]
$L_P$ = path loss [dB]

### 2.3.2. Multipath and Fading

In our increasingly mobile lifestyles, multipath and fading have become important challenges faced by most of today's wireless applications. Fading refers to fluctuation of the RF signal amplitude at the receiver antenna over a small period of time. Fading is caused by interference between various versions of the same RF signal that arrive at the receiver antenna at different times. These versions of the RF signal, which are called multipath waves, have taken different paths during their propagation and been subjected to different phase shift and amplitude attenuation. They may even be the subject of a Doppler shift caused by a mobile object or antenna. Fig. 2.15 shows such a scenario, where multiple reflections of the same signal arrive at the receiving antenna at different phase and amplitude and with a Doppler frequency shift. A simple formula describing the impact of a moving receiver or a transmitter is shown below.

$$F_D = \text{Doppler shift} = V/\lambda \text{ [Hz]} \tag{2.18}$$

where
$F_R$ = received frequency [Hz]
$F_T$ = transmitted frequency [Hz]
$V$ = relative velocity [m/s]

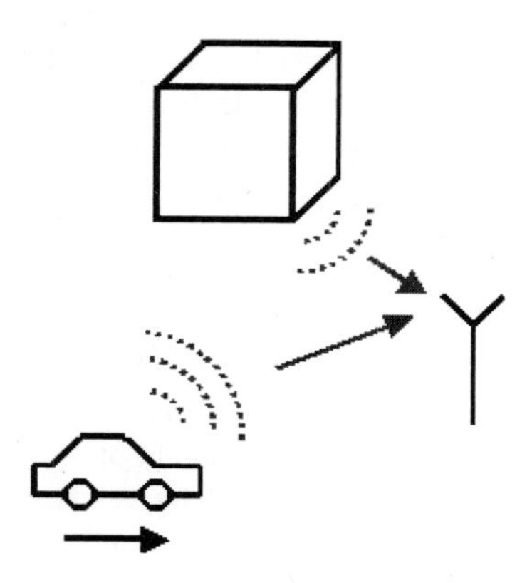

**Fig. 2.15.** Figure demonstrating the multipathing phenomenon.

Formulating fading and multipathing is somewhat more involved and is covered in detail in the literature [3].

Several techniques have been developed to help combat fading and to improve the general link performance of a system in a hostile environment. Some of the more popular techniques are discussed in the rest of this section.

### 2.3.3. Equalization

Equalization is used to combat intersymbol interference (ISI) caused by multipath within the channels. This form of interference occurs when the radio channel bandwidth truncates the signal modulation bandwidth, resulting in time spreading of modulation pulses.

To reduce ISI in a mobile environment, adaptive equalization is used to track the time-varying characteristics of the channel. Typically, a known training sequence is transmitted to characterize the channel. This information is then manipulated to calculate and set the proper filter coefficients for equalization in the receiver back-end. The data is transmitted following the training sequence and received and corrected by the equalizer. In an adaptive equalizer, the filter coefficients of the equalizer are constantly optimized to compensate for the changing radio channel [3].

### 2.3.4. Diversity

Diversity is another method that can help reduce the severity of fading. There are several diversity techniques available, such as polarization, time, and frequency diversity, but the most common diversity technique is spatial diversity. In this technique, multiple receiver antennas are strategically placed at different locations. This allows the antennas to receive different versions of the transmitted signal, thus providing the receiver with a choice of which version to use.

As shown in Fig. 2.16, there are two different methods to implement a spatial diversity receiver. In one method, one receiver path can switch between multiple antennas. This requires the receiver to first test each antenna and then make a decision. In the second method, multiple independent receiver paths with their own antennas are used. In this case, the receiver back-end can have both signals available simultaneously while choosing the better one. This allows for more dynamic operation but the cost of the receiver is significantly increased.

### 2.3.5. Coding

Another method used to improve the performance of the communications link is channel coding. In this technique, redundant data bits are added to the original message prior to modulation and transmission of signal. These added bits follow specific code sequences that help the receiver to detect and correct some or all of the error created by the radio channel. The addition of coding bits does reduce the overall channel capacity but is very effective in reducing errors.

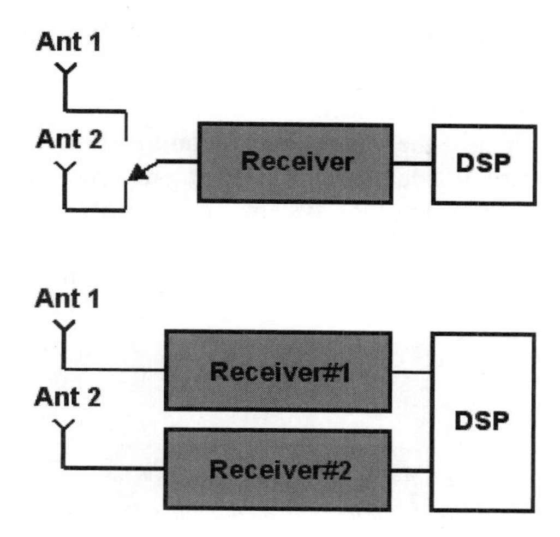


**Fig. 2.16.** Block diagram of a receiver using antenna diversity, and another one using a complete receive-chain diversity.

## 2.4. INTERFACE PLANNING

For a working communication link, a receiver should be able to effectively interface with its environment and other components in the link. This should be addressed by interface planning during the system-level design of the receiver and prior to determining the block-level specification. Interface planning helps to determine the number of ICs in a chipset and their input and output characteristics. The perceived nature of signal at different points in the receiver path and at input and output pins of the ICs is also determined. This can vary between power at RF, voltage at IF and baseband, and current at the input to an analog-to-digital converter. All of these factors will significantly impact the design and implementation of the IC(s) in the later stage of development.

In most modern receivers, the level of on-chip integration is continuously growing to include components that operate at RF, IF, and baseband frequencies. The traditional impedance used for the RF components is 50 ohms, but this impedance typically changes to 200 ohms at IF and higher at baseband. In designs that cover receiver blocks from RF to baseband, it is recommended to use a voltage method for tracking power, noise, and intermodulations throughout the system.

## 2.5. CONCLUSION

In this chapter we have introduced readers to the system design aspects of receiver subsystems. This is the foundation for the circuits and systems that will be discussed in later chapters. This basic foundation of receiver system design will help readers to better understand the reasons behind various performance criteria placed on individual components of the receiver.

Additionally, in light of the current integration trends that have taken over the semiconductor industry, it is highly unlikely that a designer can work successfully without a basic understanding of the system. The majority of receiver ICs have already evolved into integrated subsystems and a receiver designer can no longer limit his/her understanding or responsibility to a single component in the receiver chain.

## REFERENCES

1. T. Pratt and C. W. Bostian, *Satellite Communications,* Wiley, New York, 1986.
2. L. W. Couch II, *Digital and Analog Communication Systems,* Macmillan, New York, 1993.
3. T. S. Rappaport, *Wireless Communications Principles and Practice,* Prentice-Hall, Upper Saddle River, 1996.

# 3

# REVIEW OF RECEIVER ARCHITECTURES

## INTRODUCTION

In this chapter, we review different RF front-end receiver architectures with the advantages and disadvantages associated with each of them. The two broad categories of receivers are IF and zero IF. In common terms, these are known as heterodyne and homodyne receivers, respectively. As the radio front-ends continue to evolve toward more on-chip integration, IF frequency continue to decrease, giving rise to different architectures: high IF, very low IF, and zero IF. These are broadly categorized under heterodyne and homodyne, depending on the frequency planning. In this chapter, we discuss in detail the architecture and specific issues involved in the development of receiver front-ends, from a system level. First, we present the specific architectures at the block level and discuss their operation, and then take the reader to an in-depth understanding of the various issues involved in the system.

Heterodyne receiver topologies are well known and widely utilized for current wireless applications. In such topologies, the input RF signal is down-converted to an intermediate frequency (IF) where it is amplified and filtered before the final demodulation by a low-frequency demodulator [1–3]. Typically, this demodulator is built to operate at frequencies below 100 MHz; therefore, two intermediate conversions are sometimes needed to facilitate image filtering by using an additional IF sufficiently different from the RF signal. Consequently, these multiple stages of filtering and amplification add to the complexity and cost of the receiver.

Two fundamental operations of a receiver include down-conversion and demodulation. In the down-conversion function, the wanted signal is filtered and separated from the interferers, and converted from the carrier frequency to a frequency suitable for the demodulator. Demodulation is performed at lower frequency, either by

*Modern Receiver Front-Ends.* By J. Laskar, B. Matinpour, and S. Chakraborty
ISBN 0-471-22591-6 

a simple in-phase and quadrature phase (I/Q) demodulator, or digitally sampled and performed by a digital signal processor (DSP). The latter allows for the use of complicated modulation schemes and complex demodulation algorithms.

The main focus of modern communication devices is to provide more integration on-chip. The use of lower IF or elimination of IF from the frequency plan has many implications for the receiver architecture. Low IF receivers combine the advantages of zero IF and IF architectures. They can achieve the performance advantages of IF receivers, achieving the high level of integration found in zero IF receivers. A detailed illustration of the architectures is presented in [2].

## 3.1. HETERODYNE RECEIVERS

Heterodyne receivers have been in use for a long time and have been quite popular since the early days of radio communication. Figure 3.1 provides a schematic representation of a heterodyne receiver. One or two stages of down-conversion can be used in this architecture. One major issue in heterodyne receivers is the suppression of unwanted image signals. One method to perform image rejection is to provide a high-frequency front-end filter with a very high quality factor ($Q$). It is very hard to integrate such a filter, and the existing solutions are quite expensive, as well as vulnerable.

The frequency planning for the heterodyne receiver is shown in Fig. 3.2 in the down-conversion path. The desired signal is situated at a frequency of $f_c$, and the image frequency is at $(f_c - 2f_i)$. The mixer produces frequency components on $f_i$ for $f_x = f_c$ and $f_x = f_c - 2f_i$. Representing the antenna signal $a(t)$ by a phase-modulated signal as $\cos[2\pi f_x t + m(t) + \phi]$, multiplication by a sinusoidal waveform yields

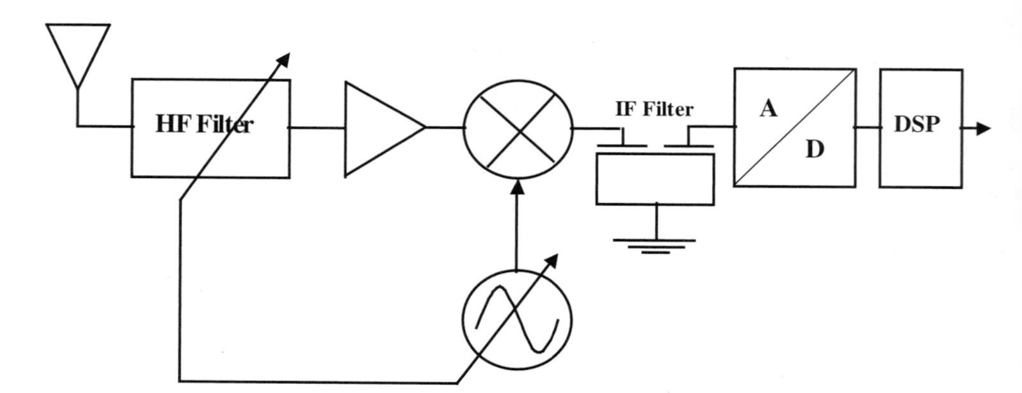


**Fig. 3.1.** A general heterodyne receiver architecture.

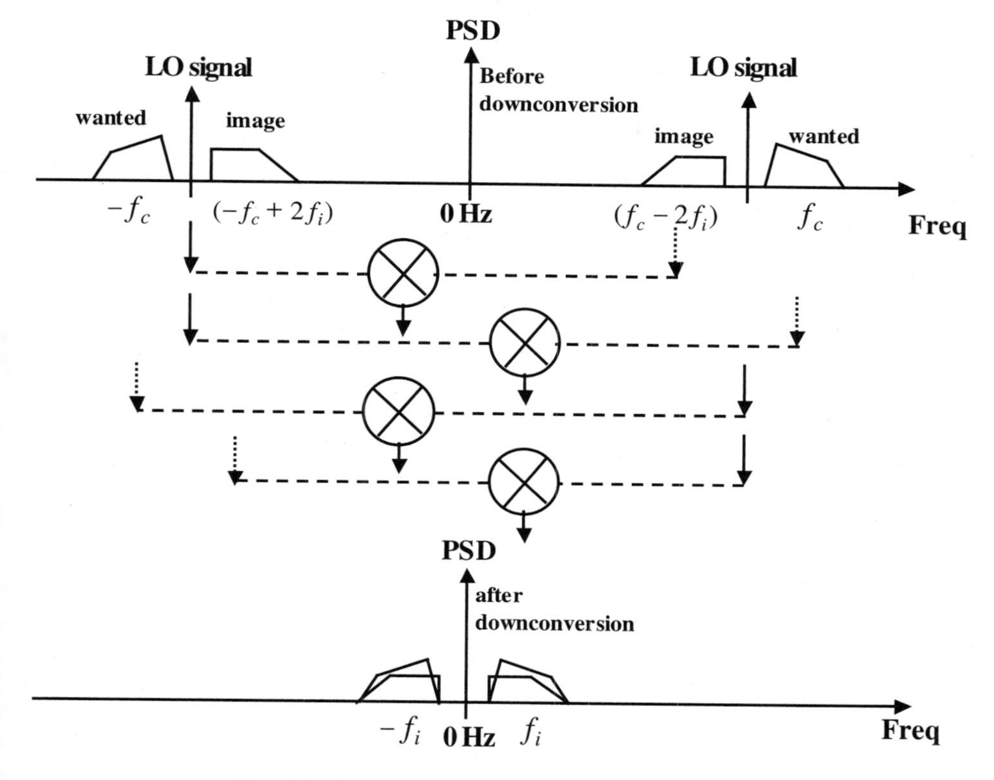


**Fig. 3.2.** Frequency scheme in a heterodyne architure.

$$
\begin{aligned}
a(t)\sin[2\pi(f_c - f_i)t + \varphi] &= \cos[2\pi f_x t + m(t) + \phi]\sin[2\pi(f_c - f_i)t + \varphi] \\
&= \tfrac{1}{2}(\sin[2\pi(f_x + f_c - f_i)t + m(t) + \phi + \varphi] \\
&\quad - \sin[2\pi(f_x - f_c + f_i)t + m(t) + \phi - \varphi])
\end{aligned}
\tag{3.1}
$$

It is a prime concern to suppress the image signal prior to down-conversion to IF. This is usually performed by high-frequency filters. The implementation of high-frequency filters becomes easier, when $f_i$ is high enough, so that the wanted signal is relatively far away from the image frequency.

After the down-conversion to IF, the filtering must be incorporated again, and a high-order filter is usually necessary. Integration of these IF filters is also very hard and most of the existing technologies use ceramic resonators. It is also reported in the literature [4–6] that the performance of the integrated active bandpass filters (achievable dynamic range over power consumption) is quality factor ($Q$) times worse than the performance of a passive bandpass filter. Fig. 3.3 shows an illustration of a scenario in which an IF signal is demodulated using a DSP.

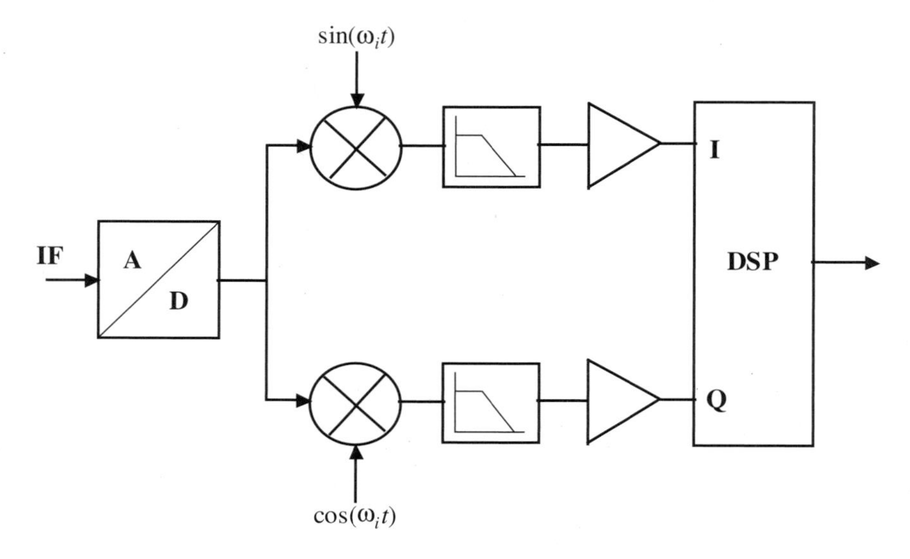


**Fig. 3.3.** Demodulation with a DSP of a signal situated on IF.

## 3.2. IMAGE REJECT RECEIVERS

One of the major drawbacks of heterodyne receiver topologies is the image frequency interference. As described earlier in Chapter 2 and shown in Eqns. 3.2 and 3.3, the image frequency band is located on the opposite side from the desired signal across from the LO signal. In the mixer, the image band is down-converted onto the IF band alongside the desired signal. Therefore, noise and signals existing in the image band interfere and corrupt the desired signal located in the IF band.

$$F_{\mathrm{RF}} = F_{\mathrm{LO}} \pm F_{\mathrm{IF}} \tag{3.2}$$

$$F_{\mathrm{image}} = F_{\mathrm{LO}} \pm F_{\mathrm{IF}} \tag{3.3}$$

To maintain an acceptable receiver signal quality, most modern wireless standards require 60 to 90 dB of image rejection. The traditional method of image rejection is the use of filters designed with a stop-band at the image frequency. However, due to the stringent requirements, image rejection is typically performed through a combination of filtering and the use of image rejection mixing techniques. Image rejection topologies are especially important for development of integrated solutions and reducing the use of off-chip components that are typically used for filters. The two most popular image rejection techniques are Hartley and Weaver architectures. These techniques, which are described in the following sections, typically yield about 30 to 35 dB of image rejection.

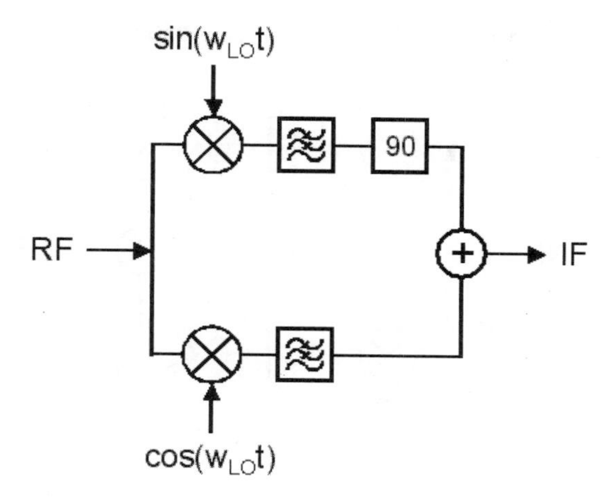


**Fig. 3.4.** Hartley image rejection architecture.

### 3.2.1. Hartley Architecture

The Hartley image rejection architecture is demonstrated in Fig. 3.4. The RF signal is down-converted by quadrature LO signals. The resulting IF signals are then low-pass-filtered and after one is phase-shifted by 90°, the IF signals are combined. In this process, depending on the IF path that is subjected to the 90° phase-shifter, either the image band or the receive band is cancelled. The mathematical details of Hartley image rejection are described in the literature [1].

A major problem with the Hartley architecture is its sensitivity to phase and amplitude imbalance between the quadrature mixers or the two IF paths. The imbalance is typically caused by mismatch between the quadrature mixers, in the two IF paths, or in the phase-shifter. As shown in Eqn. 3.4, image rejection ratio (IRR) of the Hartley architecture can be described as a function of phase and amplitude imbalance.

$$IRR = \tfrac{1}{4} \times [(\Delta A/A)^2 + \theta^2] \tag{3.4}$$

where
$\Delta A/A$ = relative gain mismatch
$\theta$ = relative phase mismatch in radians

Using the relationship described above, it can be seen that an amplitude mismatch of 0.1 dB and a phase mismatch of 1 degree yields an approximate 40 dB of IRR.

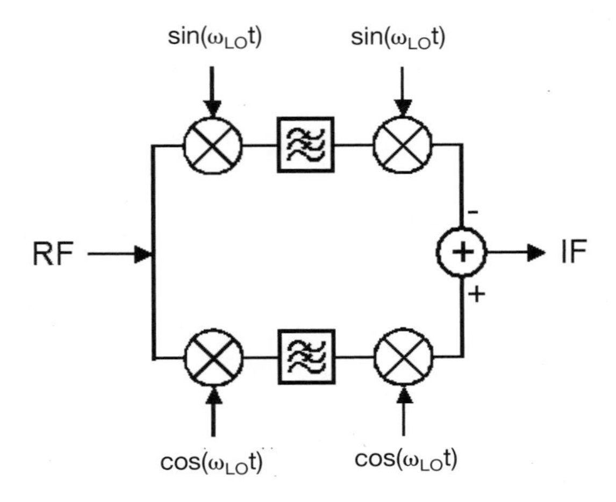


**Fig. 3.5.** Weaver image rejection architecture.

### 3.2.2. Weaver Architecture

The Weaver architecture is shown in Fig. 3.5. This architecture utilizes an additional pair of mixers to perform the phase shifting prior to IF combination. The use of the second set of mixers brings about concern for a second image, which must be addressed using proper frequency planning and filtering. Use of zero as the second IF can eliminate the second image problem.

The Weaver architecture is also sensitive to phase and amplitude imbalance and its performance can be described by Eqn. 3.4, similar to the Hartley structures. The Weaver architecture is able to achieve better image rejection because it does not suffer from the amplitude mismatch caused by the phase shifter used in the Hartley architecture.

## 3.3. ZERO IF RECEIVERS

Figure 3.6 shows a direct conversion receiver architecture. In the case of direct conversion, or zero IF receivers, the desired signal is directly down-converted to baseband. With this scheme, the image signal becomes a part of the desired signal itself. Both the desired and the image signals are mirror images of each other around the frequency axis, which results in lower and upper sidebands falling on top of each other. The frequency scheme is shown in Fig. 3.7.

This problem is taken care of by performing the down-conversion twice, with sine and cosine waveforms. The quadrature down-conversion is then calculated as:

$$u_i(t) = a(t)\cos(2\pi f_c t + \varphi),\ u_q(t) = a(t)\sin(2\pi f_c t + \varphi) \tag{3.5}$$

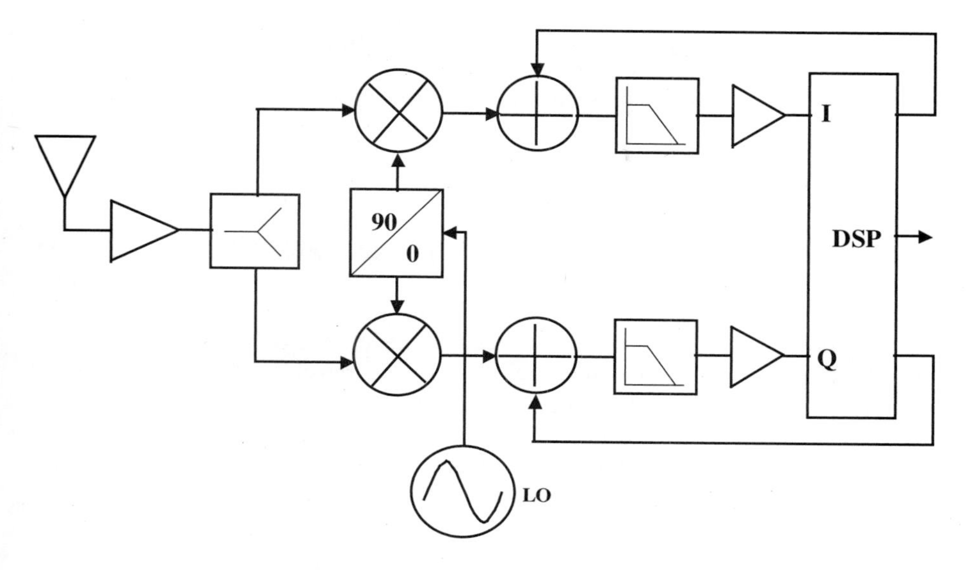


**Fig. 3.6.** Direct conversion receiver with a DC compensation feedback path.

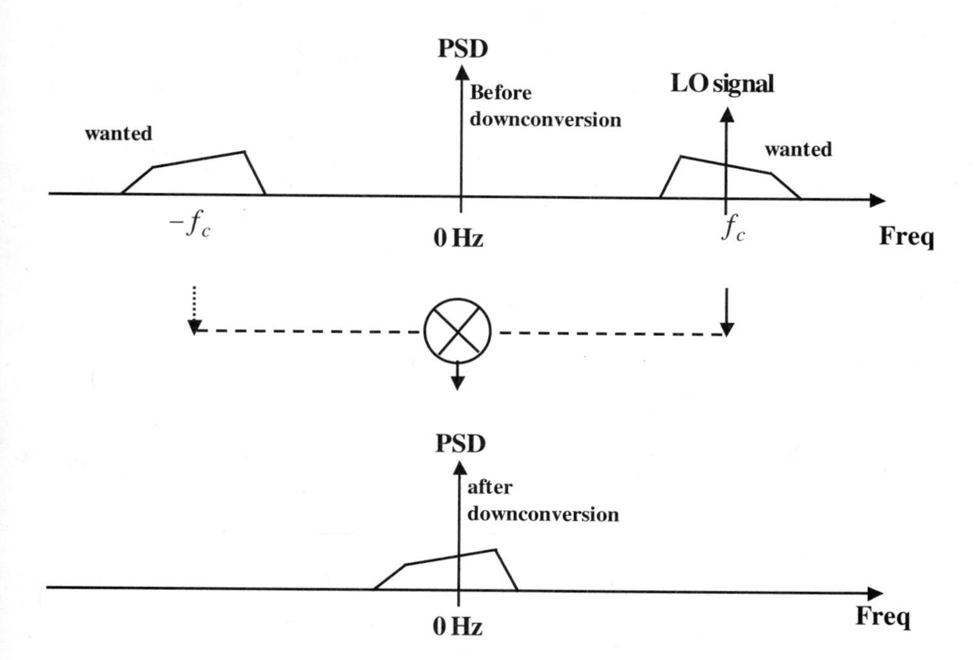


**Fig. 3.7.** Frequency scheme in DCR with respect to a single positive frequency.

Suppression of the image is no longer a concern in direct conversion receivers, as opposed to IF receivers. A detailed illustration of the issues in the case of direct conversion architectures is presented later in this chapter.

## 3.4. LOW IF RECEIVERS

Figures 3.8 and 3.9 shows the low-IF receiver architecture; its frequency planning is shown in Fig. 3.10. The development of low-IF topologies started with the target to combine the advantages of both types of receivers, IF and direct conversion. Using two down-conversion paths in the receiver, the image would still be available at two IF frequencies. Thus, the image rejection can be transferred to the low-frequency IF part. This eliminates the necessity of a very high $Q$ filter in the front-end. The IF filter becomes a very low $Q$ bandpass filter ($Q = 1$ or 2), which is quite easy to integrate on-chip. An interested reader is recommended to refer to [7] for a hardware implementation of a low-IF receiver in a silicon-based technology.

## 3.5. ISSUES IN DIRECT CONVERSION RECEIVERS

In this section, we illustrate the issues and challenges involved in implementing a direct conversion receiver. Direct conversion receivers have been proposed since the early days of radio, but the system performance has been usually poor. With the advances in the digital communication area, a lower performance can be accepted with the advantage of the highest possible on-chip integration provided by direct conversion architecture.

Direct down-conversion (DDC) topology is an alternative to the traditional heterodyne topology. It allows for the elimination of the IF stages and, hence, the com-

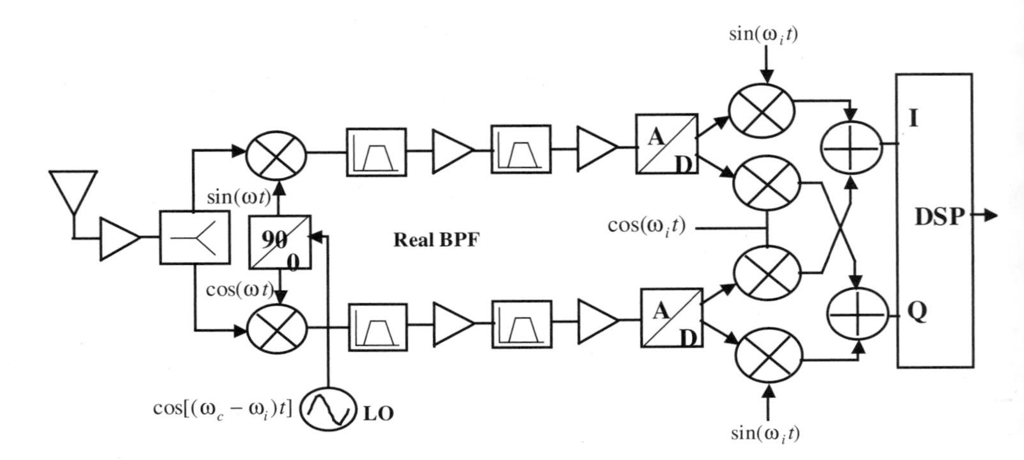


**Fig. 3.8.** Fully elaborated low-IF architecture with real filters.

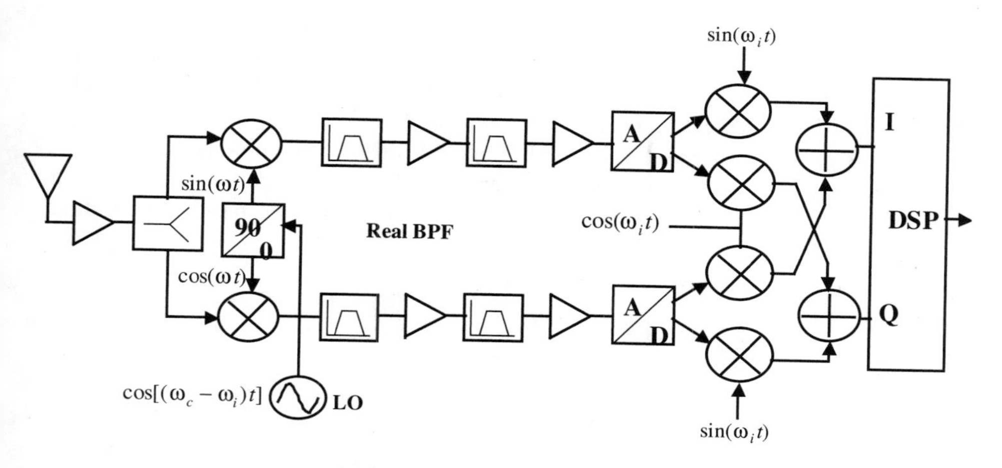


**Fig. 3.9.** Fully elaborated low-IF architecture with complex bandpass filters.

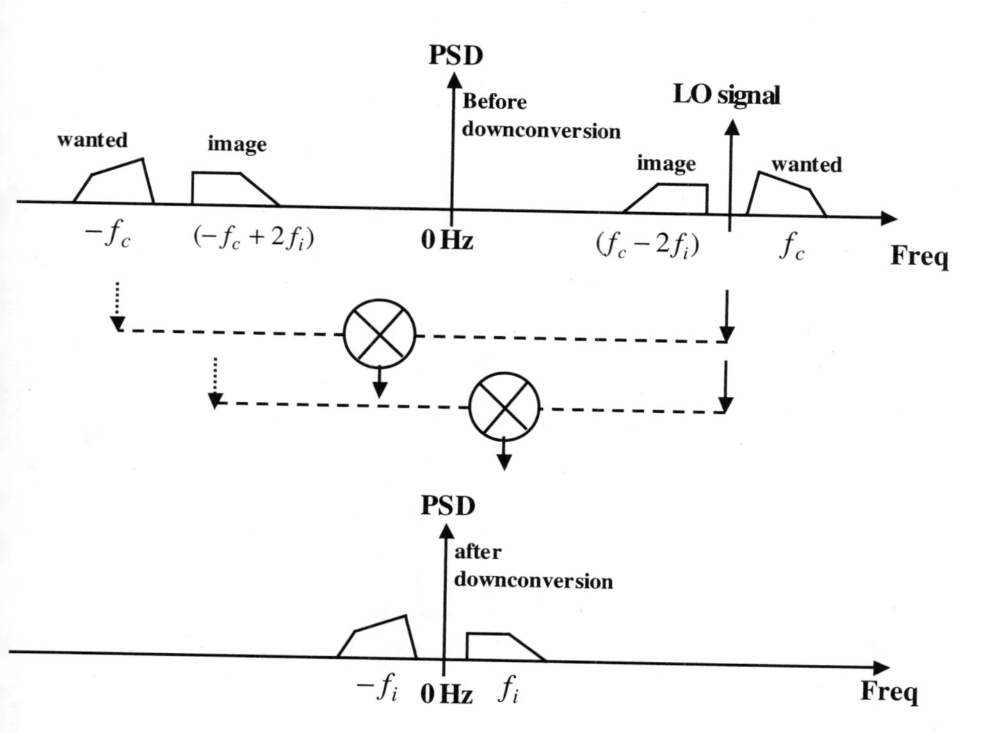


**Fig. 3.10.** Frequency planning for the low IF receiver.

plexity and cost. Essentially, a DDC receiver is a demodulator with much higher dynamic range and higher frequency of operation than a typical I/Q demodulator. Figure 3.11, demonstrates block diagrams of a typical heterodyne receiver and a DDC receiver in which the IF stage is eliminated.

In direct conversion receivers, the RF signal is directly down-converted to baseband where it can be immediately converted to digital information. As shown in Fig. 3.11, there are several options for the order in which the baseband building

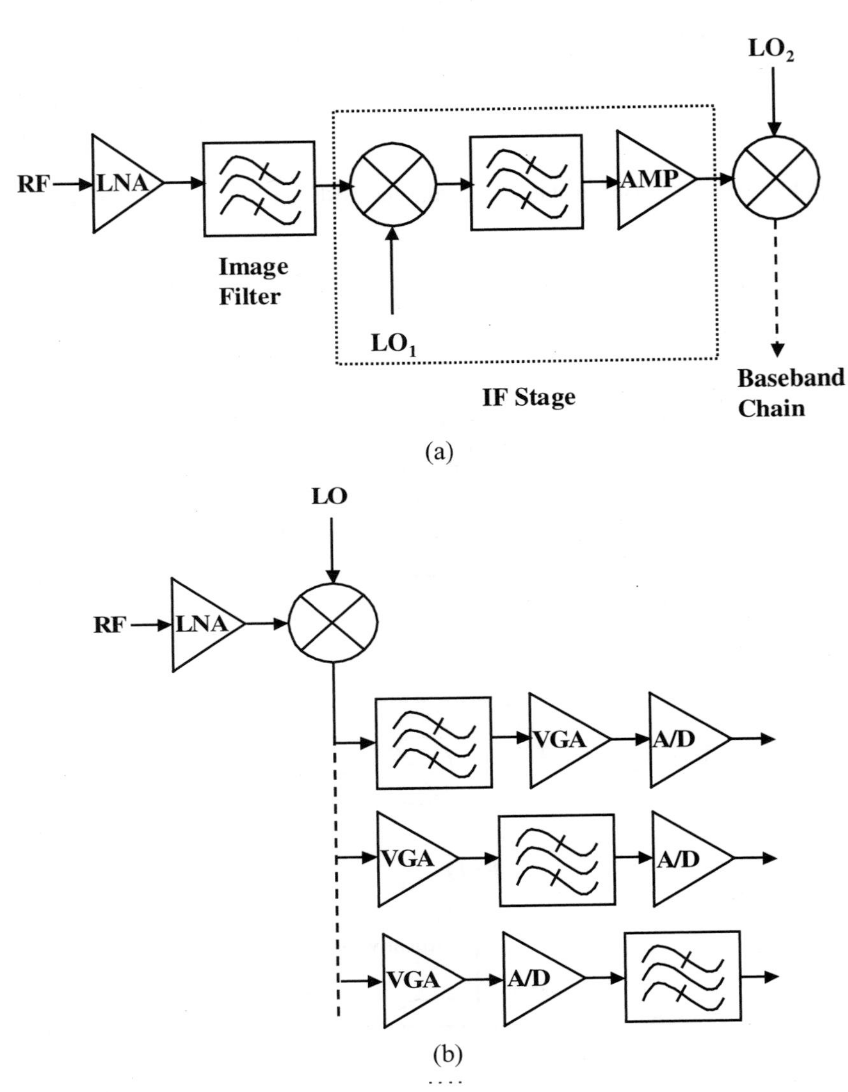


**Fig. 3.11.** Block diagram of (a) a typical heterodyne and (b) a DDC receiver.

blocks can be arranged. Depending on the chosen configuration, the linearity and noise performance of the front-end and baseband blocks can be defined.

By placing the low-pass filter (LPF) in the front, we can filter much of the extraneous signals that result from the mixing and relax the linearity requirements of the variable-gain amplifier (VGA) and the analog-to-digital converter (A/D). However, the loss through the LPF reduces the gain before the VGA and A/D and, consequently, increases the overall noise figure (NF) of the receiver. As we move the LPF further down the baseband chain, the NF improves, whereas the linearity requirements of the VGA and A/D increase. The designer will have to find an adequate balance between NF and linearity for each of the baseband blocks to maintain signal integrity. This choice often becomes a function of the semiconductor technology used for the physical implementation of the IC.

Figure 3.12 shows the frequency schemes of DDC and heterodyne receivers and highlights the intermodulation distortions and other interfering signals that are of concern for each topology. In direct conversion receivers, the higher-frequency components of the output can be easily removed by active or passive low-pass filtering in the baseband. However, because of the presence of the DC offsets and second- (IM2) and third-order intermodulations (IM3), the separation of the down-converted output from all of the extraneous signals becomes troublesome. In the heterodyne approach, however, DC offsets and IM2 can be easily eliminated through simple capacitive coupling, but the image interference and IM3 remain as in-band signals. The major design considerations for a DDC receiver can be sum-

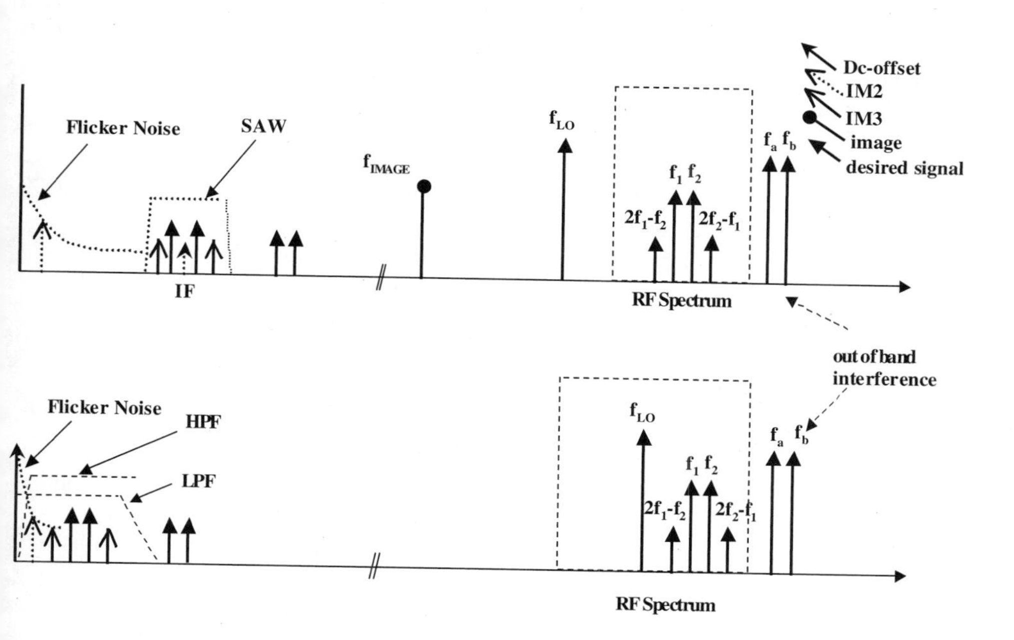


**Fig. 3.12.** Frequency scheme of heterodyne (top) and DDC (bottom) receivers.

marized as noise, LO leakage and radiation, I/Q imbalance, DC offsets, and intermodulations.

### 3.5.1. Noise

Noise performance in DDC receivers is evaluated in two distinct ways. One is the typical additive noise described by the noise figure and the other is the baseband low-frequency flicker noise [15, 16]. Since the signal is directly down-converted to baseband, flicker noise becomes an in-band phenomenon that needs to be considered. Fig. 3.12 shows the coexistence of the demodulated signal and the flicker noise in the baseband output of a DDC receiver.

Flicker noise of each individual transistor in the baseband circuitry contributes to the overall flicker noise effect; therefore, reducing the number of active devices in the baseband can help to minimize this. Higher-gain front-end circuitry can also help further reduce the impact of flicker noise on signal integrity by boosting the RF signal; however, this may compromise the overall linearity of the receiver. The choice of semiconductor process also impacts the level of flicker noise associated with each active element. Compound semiconductor and silicon (Si) bipolar processes typically have a lower flicker noise corner frequency than the popular CMOS devices [17].

### 3.5.2. LO Leakage and Radiation

Since a typical DDC receiver requires a LO signal frequency identical to the RF input carrier frequency, the LO signal is considered an in-band interference. This becomes more of an issue when we consider the high levels of LO power that are typically used in high-linearity mixers. This LO can couple into the antenna and not only radiate out in the receive band of other users but also penetrate and saturate the RF front-end [3]. Most popular wireless standards only allow in-band LO radiations of less than –80 dBm, which would require 80 to 90 dB isolation between the LO signal source and the antenna.

High reverse isolation in the front-end and good shielding of the receiver can reduce the LO leakage and radiation; however, alternative schemes such as subharmonic mixing can also be used to simplify this problem by moving the LO signal out of the RF frequency band of interest [18].

### 3.5.3. Phase and Amplitude Imbalance

Most modern wireless modulation scheme requires $I$ and $Q$ signal separation and demodulation to fully recover the information. As shown in Fig. 3.13, this is accomplished by using two down-converters operating in parallel with 90° of phase differentiation [15].

Unfortunately, implementing accurate phase shifters at higher frequency becomes a challenging task. Heterodyne receivers perform phase differentiation at much lower IF frequencies through simple polyphase circuitry with great accuracy.

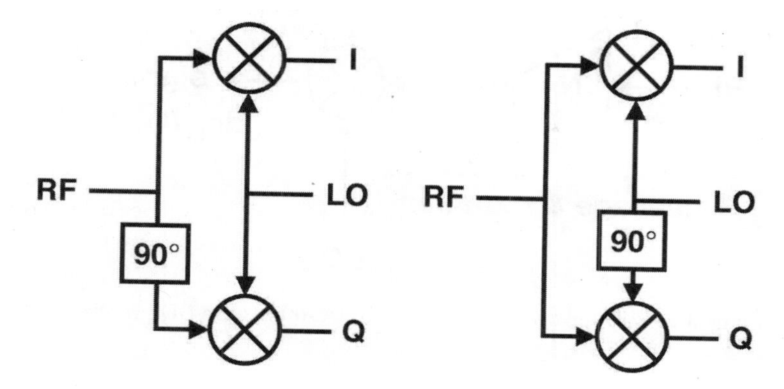


**Fig. 3.13.** *I*/*Q* demodulator with the phase shifter on (a) the RF and (b) the LO path.

Phase shifters used at higher frequencies are either too lossy or suffer from phase and amplitude imbalance generated by parasitics.

For a received signal defined as $V_{\text{in}} = I(t)\cos(\omega_{RF}t) + Q(t)\cos(\omega_{RF}t)$, and amplitude and quadrature phase imbalance of $\varepsilon$ and $\theta$, respectively, we can describe the baseband $I$ and $Q$ voltages as

$$V_I = I(t)(1 + \varepsilon/2)\cos(\theta/2) - Q(t)(1 + \varepsilon/2)\sin(\theta/2) \tag{3.6}$$

$$V_Q = I(t)(1 - \varepsilon/2)\sin(\theta/2) + Q(t)(1 - \varepsilon/2)\cos(\theta/2) \tag{3.7}$$

From these equations, we deduce that $\varepsilon$ appears as a nonunity scale factor in the amplitude, whereas $\theta$ results in cross talk between the demodulated $I$ and $Q$ waveforms, degrading the signal-to-noise ratio (SNR). In practice, for SNR degradation of less than 1 dB in a QPSK signal, $\varepsilon < 1$ and $\theta < 1$ dB are required.

### 3.5.4. DC Offset

Extraneous DC voltages in the demodulated spectrum of a DDC receiver not only corrupt the output, but also propagate through the baseband circuitry and saturate the subsequent stages. These DC offsets are mostly generated through self-mixing the LO signal and mismatch in the mixers [1–3].

In direct conversion receivers, the mixer is immediately followed by LPFs and a chain of high-gain direct-coupled amplifiers that can amplify small levels of DC offset and saturate the stages that follow. Consequently, sensitivity of the receiver can be directly limited by the DC offset component of the mixer output. The DC offset of a mixer can be separated into two components, a constant and a time-varying offset. The constant DC offset can be attributed to the mismatch between mixer components while the time-varying DC offset is generated by self-mixing of the LO. As demonstrated in Fig. 3.14, because of an imperfect isolation between the LO and the RF ports of the mixer, a finite amount of LO leakage into

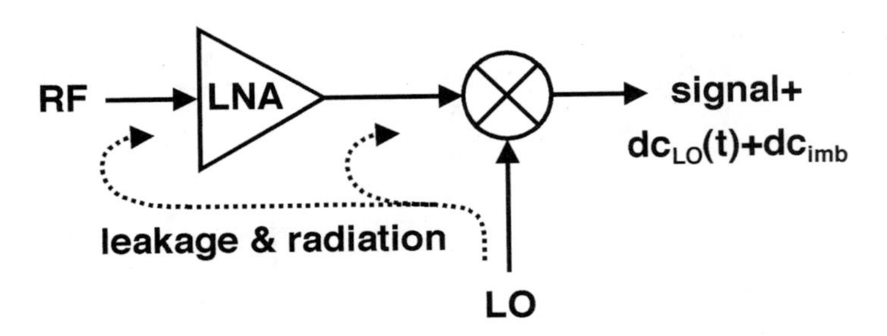


**Fig. 3.14.** LO leakage and generation of DC offsets in a DDC receiver.

the RF port persists. In addition, LO leakage and even the LO radiation can reach the low-noise amplifier (LNA) or other stages prior to the mixer and propagate through the front-end. These signals can get amplified before arriving at the mixer and bring about a serious issue of self-mixing that results in the generation of a relatively large DC component at the mixer outputs. The level of this DC offset, which is dependent on the time-varying load of the antenna, can also vary with time.

Representing LO leakage and radiation level by $C_{\mathrm{L+R}}$, mixer conversion gain by $C_{\mathrm{mixer}}$, and LO level by $C_{\mathrm{LO}}$, Eqns. 3.8 through 3.12 demonstrate the recovery of a baseband tone ($\omega_a$) by the LO frequency ($\omega_{\mathrm{LO}}$). As observed in Eqn. 3.12, the amplitude of the DC offsets generated from the self-mixing is mainly proportional to the LO leakage and radiation $C_{\mathrm{L+R}}$, mixer gain, and the LO power level $C_{\mathrm{LO}}$.

$$\omega_1 = \omega_{\mathrm{RF}} + \omega_a \text{ (desired signal)} \tag{3.8}$$

$$\omega_{\mathrm{LO}} = \omega_{\mathrm{RF}} \text{ (for direct conversion)} \tag{3.9}$$

$$RF_{\mathrm{input}} = \cos(\omega t - \omega_1 t) + C_{L+R}\cos(\omega t - \omega_{\mathrm{LO}} t) + \ldots \tag{3.10}$$

$$LO = C_{\mathrm{LO}}\cos(\omega t - \omega_{\mathrm{LO}} t) \tag{3.11}$$

$$IF = RF_{\mathrm{input}} \times LO = \tfrac{1}{2} C_{\mathrm{mixer}} C_{\mathrm{LO}}[\cos(\omega t - \omega_a t) + C_{\mathrm{LO}} C_{L+R} + \ldots] \tag{3.12}$$

A possible solution for removing the DC offset is AC coupling of the mixer output. This will not only remove the unwanted DC offsets, but at the same time will corrupt the down-converted signal by attenuating the components near DC. Unfortunately, this is not acceptable for demodulating most random binary modulation schemes that exhibit a DC peak in their signal spectrum. Use of "DC-free" modulation schemes such as binary-frequency shift keying (BFSK) can reduce susceptibility to DC offsets while taking away spectrum efficiency and other advantages of popular and mainstream digital modulation schemes. Therefore, DC offset cancellation techniques are necessary to accommodate the use of direct conversion topology in today's wireless applications. The signal distortion would be lower and could be of an acceptable level, if the time constant is longer (~ 1s), but this long time constant makes the settling time of the receivers too long, and cannot be integrated as a part of analog circuits [2].

Many DC offset cancellation techniques have been reported over the past few years. They can be divided to baseband analog and digital techniques [1–3, 8, 19–21]. Most of the reported analog techniques require large off-chip capacitors, and the digital solutions require complex digital circuitry, which is typically implemented in a separate chip. In the DSP, a complex nonlinear scheme can be used toward estimation of DC offset in a dynamic manner. In many cases, the reported techniques only provide solutions for specific system topologies.

### 3.5.5. Intermodulations

Direct down-conversion receivers are not only susceptible to both odd- and even-order intermodulations, but also vulnerable to the even orders. The mixing equations from Section 3.5.4 are now modified and represented in Eqns. 3.13 through 3.19 to incorporate two-tone mixing and generation of IM2 and IM3 signals in addition to the DC offsets. These equations show demodulation of two baseband tones, $\omega_a$ and $\omega_b$, that have been modulated on an RF carrier labeled as $\omega_1$ and $\omega_2$.

$$\omega_1 = \omega_{\mathrm{RF}} + \omega_a \tag{3.13}$$

$$\omega_2 = \omega_{\mathrm{RF}} + \omega_b \tag{3.14}$$

$$\omega_{\mathrm{LO}} = \omega_{\mathrm{RF}} \text{ (for direct conversion)} \tag{3.15}$$

$$\begin{aligned} RF_{\mathrm{input}} = {} & C_1\cos(\omega t - \omega_1 t) + C_2\cos(\omega t - \omega_2 t) + C_{\mathrm{IM3}}\cos[\omega t - (2\omega_2 - \omega_1)t] \\ & C_{\mathrm{IM3}}\cos[2\omega t - (2\omega_1 - \omega_2)t] + C_{\mathrm{L+R}}\cos(\omega t - \omega_{\mathrm{LO}} t) + \ldots \end{aligned} \tag{3.16}$$

$$LO = C_{\mathrm{LO}}\cos(\omega t - \omega_{\mathrm{LO}} t) \tag{3.17}$$

$$IF = RF_{\mathrm{input}} \times LO \tag{3.18}$$

$$IF = \tfrac{1}{2} C_{\mathrm{mixer}} C_{\mathrm{LO}} \begin{bmatrix} C_1\cos(\omega t - \omega_a t) + C_2\cos(\omega t - \omega_b t) + \ldots \\ C_{\mathrm{IM3}}\cos[\omega t - (2\omega_a - \omega_b)t] + C_{\mathrm{IM3}}\cos[\omega t - (2\omega_b - \omega_a)t] + \ldots \\ C_{\mathrm{IM2}}\cos[\omega t - (2\omega_a - \omega_b)t] + \ldots \\ C_{\mathrm{L+R}} + \ldots \end{bmatrix} \tag{3.19}$$

There are several mechanisms that can contribute to intermodulation distortions in a DDC receiver. As demonstrated in Eqn. 3.19, the major portion of the even-order intermodulations are generated by the nonlinearities in the mixer, in which two adjacent RF tones or interference signals are mixed together to generate a low-frequency beat at the mixer output. Low-noise-amplifier (LNA) nonlinearities can also generate a form of even-order distortion by similarly producing a low-frequency beat at the LNA output. After some finite attenuation by capacitive coupling, this low-frequency beat can propagate through the mixer and appear at the output port. One form of even-order intermodulation is generated by an amplitude-modulated in-band interferer. As shown in Fig. 3.15, the AM modulation on such an interferer can be demodulated in the mixer, resulting in AM noise in the baseband output.

Another source of second-order distortion is AM modulation caused by the mechanical vibrations that occur during the use of a portable product. The relationship of such vibrations to the generation of IM2 distortions are very difficult to quantify theoretically; therefore, in an attempt to delegate risk of receiver failure, designers tend to place more stringent requirements on this performance criterion.

Also demonstrated in Eqn. 3.19, the RF signal carries third-order distortions that are generated by the nonlinearities in both the LNA and mixer. These odd-order distortions are then demodulated by the LO signal and down-converted to the baseband, thus adding to the additional odd-order products generated in the mixer, further corrupting the desired signal.

Balanced topologies have previously proven to be effective in reducing intermodulations and DC offsets in lower RF frequencies [10, 11]. However, implementation of balanced structures at higher frequencies is much more difficult because of the higher level of process variations. In particular, technologies such as compound semiconductor metal–semiconductor field effect transistor (MESFET) processes, which are more suitable for high-frequency applications, suffer from a low level of device conformity. This variation is less pronounced in diodes than in field effect transistor (FET) structures, therefore allowing for implementation of reasonably matched balanced mixers. Diode mixers have been exploited beyond microwave frequencies into millimeter wave and have proven their suitability for high-frequency mixing.

Despite using balanced topologies, the unwanted terms comprised mostly of even-order products cannot be fully suppressed and remain as barriers in achieving high dynamic range. An attractive and promising alternative is even-harmonic (EH)

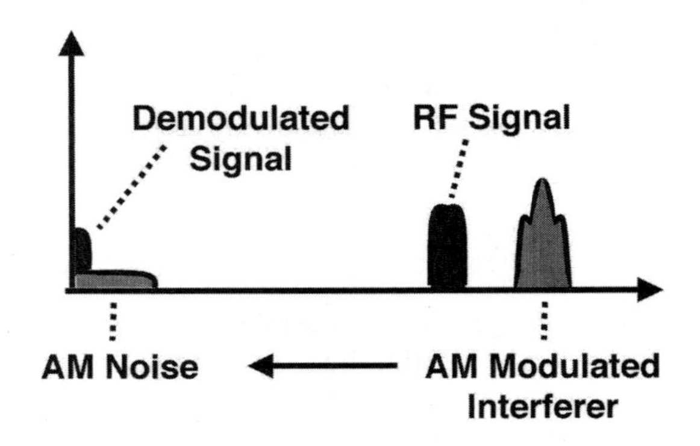


**Fig. 3.15.** Translation of AM noise in the mixer.

mixing [10, 11]. This topology has been previously reported as a suitable method for reducing the even-order products and can be used for similar purpose in a DDC receiver. Several methods of even-harmonic mixing have been introduced and implemented for DDC receivers. Even-harmonic mixing using antiparallel diode pair (APDP) mixers has been demonstrated as a feasible method to implement DDC receivers in the L-band and demonstrated a virtual elimination of even-order intermodulations [12, 13].

## 3.6. ARCHITECTURE COMPARISON AND TRADE-OFF

A detailed illustration of the architectural trade-off has been presented in [2]. The analysis uses representation of various signals in a polyphase manner. An interested reader is recommended to refer to [2] for a detailed mathematical analysis on receiver performance using polyphase signal representation of signals. In this subsection, we introduce the reader to the many trade-offs present between a low-IF and zero-IF receivers.

A major disadvantage of direct conversion receivers is the presence of DC offset, which also includes LO-to-RF cross talk. This problem is not present in low-IF receivers, as the results of cross talk is usually situated around DC, with a very small bandwidth (< 1 kHz). This is true for the product of LO multiplied with itself. The product of RF multiplied by itself is broadband, but most of its power is situated within 200 kHz. Hence, low-IF architecture is advantageous in this regard. However, with modern communication technologies, and the capabilities of "DC-free" modulation schemes, a majority of the DC offset problems do not seem prohibitive.

One disadvantage of low-IF receivers is the requirement of high levels of matching accuracy between components. The suppression of the image must be higher. In some cases, the image signal can be higher than the desired signal. Choice of the IF is very important, and plays a crucial role in terms of achievable suppression of the image signal. The matching accuracy also impacts the amount of cross talk between the complex building blocks. Another matching problem in the front-end is the phase error introduced by the quadrature oscillator.

## 3.7. CONCLUSION

In this chapter, we have introduced the reader to various front-end receiver architectures and their trade-offs. Both the low-IF and direct conversion receivers use quadrature down-conversion in the front-end. As modern communication systems evolve toward low power and more on-chip integration, it would be quite interesting to see which receiver architecture would be suitable for implementation. Direct conversion receivers become quite attractive in the prospect of a "DC-free" modulation scheme. However, selection of a particular topology depends entirely on the application in perspective. As discussed in the previous sections, along with the system architecture and frequency planning, a proper choice of hardware and careful integration is also important.

## REFERENCES

1. A. A. Abidi, "Direct-conversion radio transceivers for digital communications," *IEEE Journal of Solid-State Circuits, 30,* 12, 1399–1410, Dec. 1995.
2. J. Crols and M. S. J. Steyaert, "Low-IF topologies for high-performance analog front-ends of fully integrated receivers," *IEEE Transactions on Circuit and Systems—II: Analog and Digital Signal Processing, 45,* 3, March 1998.
3. B. Razavi, "Design considerations for direct-conversion receivers," *IEEE Transactions on Circuit and Systems, 44,* 6, 428–435, June 1997.
4. Y.-T. Wang, F. Lu, and A. A. Abidi, "CMOS active filter design at very high frequencies," *IEEE Journal of Solid-State Circuits,* pp. 1562–1574, Dec 1990.
5. G. Groenwold, "The design of high dynamic range continuous-time integrable bandpass filters," *IEEE Transactions on Circuit and Systems, 38,* 838–852, August 1991.
6. J. O. Voorman, "Continuous-time analog integrated filters," in *Integrated Continuous-Time Filters,* IEEE Press, New York, pp. 27–29. 1993.
7. F. Behbahani, J. C. Leete, Y. Kishigami, A. Roithmeier, K. Hoshino, and A. A. Abidi, "A 2.4-GHz low-IF receiver for wideband WLAN in 0.6-μm CMOS—architecture and front-end," *IEEE Journal of Solid-State Circuits, 35,* 12, December 2000.
8. A. Rofougaran, G. Chang, J. J. Rael, Y.-C. Chang, M. Rofougaran, P. J. Chang, M. Djafari, M.-K. Ku, E. W. Roth, A. A. Abidi, and H. Samueli, "A single-chip 900-MHz spread-spectrum wireless transceiver in 1—μm CMOS—Part II: receiver design," *IEEE Journal of Solid-State Circuits, 33,* 4, 535–547, April 1998.
9. K. Itoh, M. Shimozawa, N. Suematsu, and O. Ishida, "Even harmonic type direct conversion receiver ICs for mobile handsets: Design challenges and solutions," in *IEEE RF IC Symposium,* pp. 53–56, June 1999.
10. K. Kawakami, K. Tajima, M. Shimozawa, K. Itoh, N. Kasai, and A. Iida, "Fully monolithic integrated even harmonic quadrature ring mixer with an active matched 90 degree power divider for direct conversion receivers," in *IEEE Microwave Theory and Technique Symposium,* pp. 657–660, June 1997.
11. A. Rofougaran, J. Y.-C. Chang, M. Rofougaran, and A. A. Abidi, "A 1 GHz CMOS RF front-end IC for a direct-conversion wireless receiver," *IEEE Journal of Solid-State Circuits, 31,* 7, 880–889, July 1996.
12. M. Shimozawa, K Kawakami, K Itoh, A. Iida, and O. Ishida, "A novel subharmonic pumping direct conversion receiver with high instantaneous dynamic range," in *IEEE Microwave Theory and Technique Symposium,* pp. 819–822, June 1996.
13. M. Shimozawa, K Kawakami, H. Ikematsu, K. Itoh, N. Kasai, Y. Isota, and O. Ishida, "A monolithic even harmonic quadrature mixer using a balanced type 90 degree phase shifter for direct conversion receivers," in *IEEE Microwave Theory and Technique Symposium,* pp. 175–178, June 1998.
14. C. D. Hull, J. L. Tham, and R. R. Chu, "A direct-conversion receiver for 900 MHz (ISM band) spread-spectrum digital cordless telephone," *IEEE Journal of Solid-State Circuits, 31,* 12, 1955–1963, Dec. 1996.

15. L. W. Couch, *Digital and Analog Communication Systems,* Macmillan, New York, 1993.
16. M. J. Buckingham, *Noise in Electronic Devices and Systems,* Halsted Press, New York 1983.
17. C. A. Vergers, *Handbook of Electrical Noise: Measurement and Technology,* TAB Professional and Reference Books, 1987.
18. S. A. Mass, *Microwave Mixers,* Artech House, Boston, 1993.
19. H. Yoshida, H. Tsurumi, and Y. Suzuki, "DC offset canceller in a direct conversion receiver for QPSK reception," in *IEEE Symposium on Personal, Indoor and Mobile Radio Communications,* vol. 3, pp. 1314–1318, Sept. 1998.
20. W. Bing, H. M. Kwon, and J. Mittel, "Simple DC removers for digital FM direct-conversion receiver," in *IEEE Vehicular Technology Conference,* vol. 2, pp. 1222–1226, May 1999.
21. S. Sampei and K. Feher, "Adaptive DC-offset compensation algorithm for burst mode operated direct conversion receivers," in *IEEE Vehicular Technology Conference,* vol. 1, pp. 93–96, May 1992.

# 4

# SILICON-BASED RECEIVER DESIGN

## INTRODUCTION

Following the discussion of wireless system fundamentals and architectures in Chapters 2 and 3, we now introduce a step-by-step design procedure towards the development of high-frequency RF front-ends in a silicon-based technology. As mentioned earlier, the principles and basic operation of RF circuits have been known since 1930. Many of the receiver front-ends based on active mixer topology tend to use a Gilbert cell mixer core or its modifications, which was introduced by Barry Gilbert in 1968 [1], followed by an initial patent by H. E. Jones. Design of low-noise amplifiers (LNAs) have also been a standard practice to reduce the noise figure of the receiver front-ends. However, an innovative combination of RF building blocks in the front-end is an area which is continuously evolving, and many variations are possible based on the specific requirements of the wireless communication standards. The evolution of mobile handsets and wireless LAN standards toward high-data-rate applications, coupled with the phenomenal development in low-cost semiconductor processes have imparted tremendous momentum to the development of fully monolithic receivers. Developments in wireless communication started around 1992 with the introduction of the GSM standard. It is quite interesting to note that although many wireless standards have evolved since then, with consequent reports of various solutions from different vendors, the fundamental developments in the area of RF circuits have been rather slow.

Among the silicon-based semiconductor technologies, silicon germanium (SiGe) has proven to be a very suitable candidate for RF front-end design, after the tremendous development and popularity of gallium arsenide (GaAs). Having shown a great deal of maturity in the recent past [2], SiGe is now targeted at the cut-off frequencies ($F_t$) in the range of 200 GHz [3], and might prove quite capable of being a

*Modern Receiver Front-Ends.* By J. Laskar, B. Matinpour, and S. Chakraborty
ISBN 0-471-22591-6 © 2004 John Wiley & Sons, Inc.

strong rival to high-cost technologies like GaAs *p*-HEMT in the millimeter wave (mmW) frequencies. Another mainstream activity in silicon-based technologies has been focused toward realization of RF front-ends in CMOS-based technologies [5,6]. Standard CMOS technologies are presumably the lowest-cost technologies, and it would be quite interesting to see their potential in the future in very high frequency RF front-end designs.

This chapter is intended to provide the reader with an understanding of the system/circuit interactions within the RF front-end, and provides details of practical implementation for such receivers. First, we introduce the system design, followed by details of circuit design and layout issues. Then we illustrate the design procedure and practical implementation details of two direct conversion receivers developed with SiGe BiCMOS process technology. The target application is an orthogonal frequency division multiplexing (OFDM) standard at 5.8 GHz, with a direct conversion receiver architecture.

## 4.1. RECEIVER ARCHITECTURE AND DESIGN

### 4.1.1. System Description and Calculations

Wireless LAN standards in the 5–6 GHz band, such as IEEE 802.11a [4], have shown promise in being the next-generation standard for much higher data rates (54 Mbps or more) compared to the existing WLAN standards. Operating in the unlicensed frequency bands from 5–6 GHz (also called U-NII bands), this standard utilizes orthogonal frequency division multiplexing (OFDM) as the modulation scheme to provide flexible data rate and robust performance against intersymbol interference caused by multipath effects. However, due to the high crest factor (also referred to as peak-to-average ratio) of the OFDM-modulated signal waveform, the RF front-end blocks require much higher linearity than standard quadrature phase-shift keying (QPSK) or FM receivers.

There have been many reports of silicon-based receiver front-end design in the open literature [5, 6]. These implementations are based on a superheterodyne architecture. However, the IEEE 802.11a standard provides the capability of utilizing the zero-th carrier as a null, thus facilitating the implementation of a direct conversion architecture due to the "DC-free" modulation scheme. A direct conversion receiver architecture eliminates the need for bulky image-reject filters, resulting in compact, low-power and low-cost solutions.

### 4.1.2. Basics of OFDM

Over the last 20 years, OFDM techniques and, in particular, discrete multitone (DMT) implementation, have been used in a wide variety of applications. OFDM has been particularly successful in numerous wireless applications in which superior performance in multipath environments is desirable. Wireless receivers detect signals distorted by time and frequency-selective fading. OFDM, in conjunction with proper coding and interleaving, is a powerful technique for combating wireless channel impairments. In the wireless LAN scenario, OFDM is used for high-speed data rates. Although the absolute delay spread in this environment is low, for very high data rates it becomes large compared to a symbol interval. OFDM is preferable to the use of long equalizers. Fig. 4.1 shows a system block diagram for OFDM.

The major contribution to the OFDM complexity problem was the application of the FFT algorithm to the modulation and demodulation processes. At the same time, DSP techniques were being introduced in modems. The technique involved assembling the input information into blocks of $N$ complex numbers, one for each subchannel. An inverse FFT is performed on each block, and the resultant is transmitted serially. At the receiver, the information is recovered by performing an FFT on the received block of signal samples. This form of OFDM is referred to as discrete multitone (DMT). The spectrum of the signal on the line is identical to that of $N$ separate QAM signals at $N$ frequencies separated by the signaling rate. Each such QAM signal carries one of the original input complex numbers. The spectrum of each QAM signal is of the form $\sin(kf)/f$, with nulls at the center of the other subcarriers, as shown in Fig. 4.2. Figure 4.3 shows the OFDM spectrum.

Care must be taken to avoid the overlap of consecutive transmitted blocks. This is solved by the use of a cyclic prefix. The process of symbol transmission is straightforward if the signal is to be further modulated by a modulator with $I$ and $Q$ inputs. Otherwise, it is necessary to transmit real quantities. This can be accomplished by first appending the complex conjugate to the original input block. A $2N$-

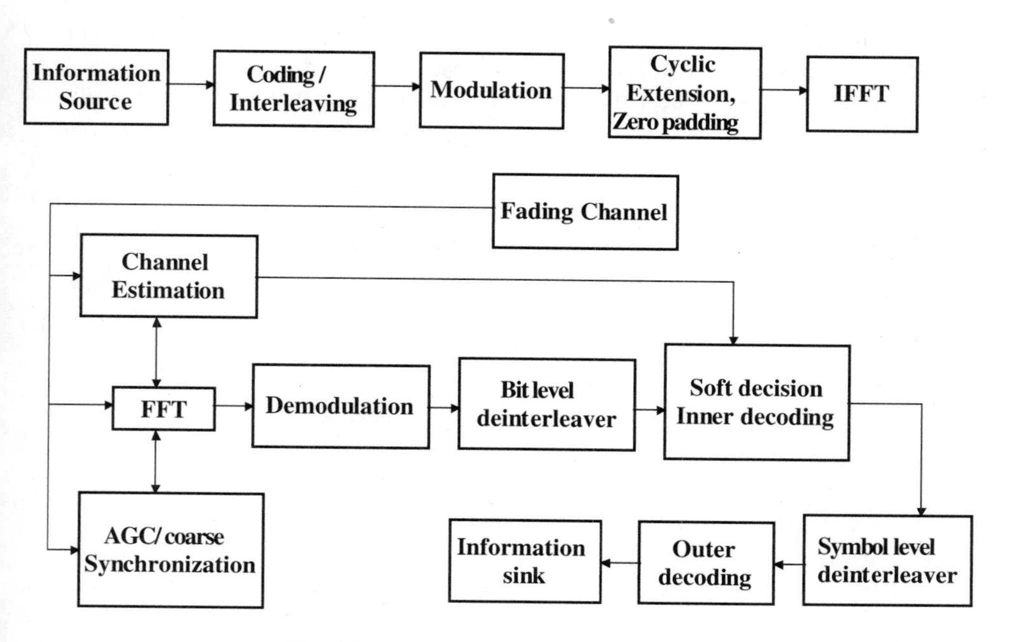


**Fig. 4.1.** Block diagram for OFDM system.

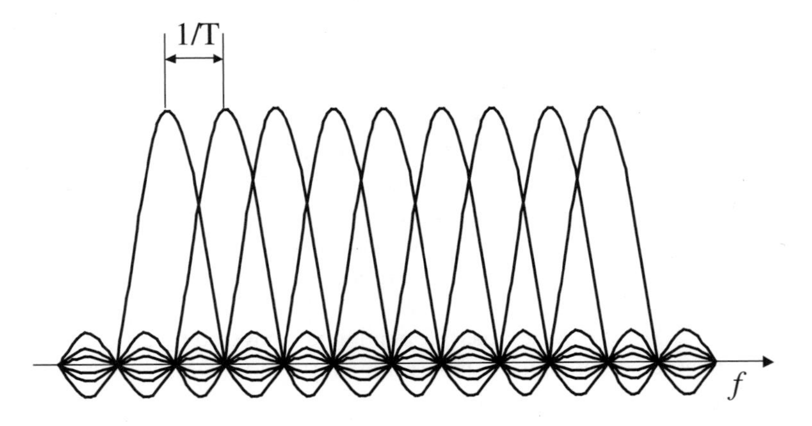


**Fig. 4.2.** Spectral overlap of subcarriers in OFDM.

point IFFT yields $2N$ real numbers to be transmitted per block, which is equivalent to $N$ complex numbers.

### 4.1.3. System Architectures

The 5–6 GHz band is divided into two parts. The lower U-NII band contains eight channels within 200 MHz (5.15–5.35 GHz) and the upper U-NII band contains four channels within 100 MHz (5.725–5.825 GHz). Figures 4.4 and 4.5 show the front-end direct conversion architectures under consideration. A 50-ohm-based approach is quite suitable and has been the conventional approach. In a 50-ohm-based receiver front-end architecture, all the individual RF building blocks are designated by an

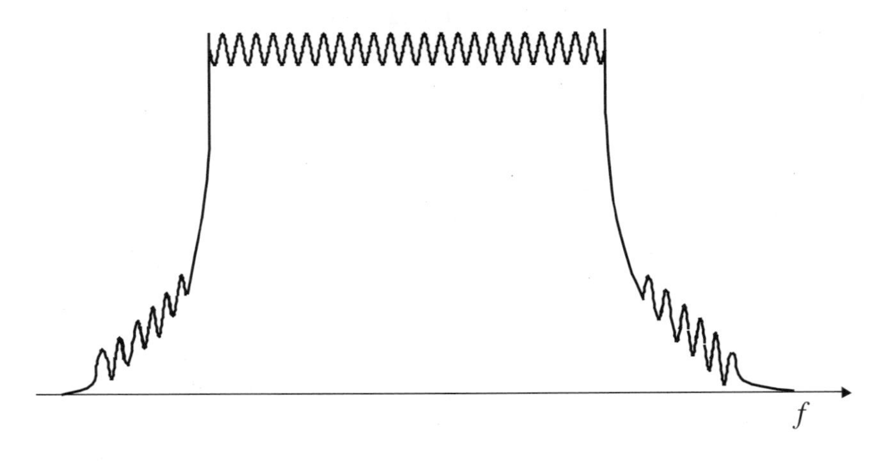


**Fig. 4.3.** Spectrum of OFDM signal.

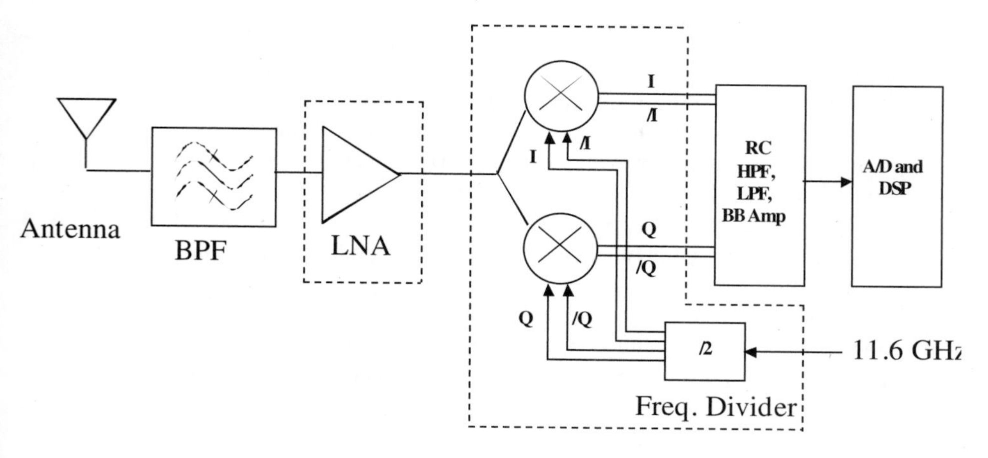


**Fig. 4.4.** Architecture I: 50 Ω receiver front-end architecture.

input and output impedance of 50 ohms. In a typical design project, it helps tremendously in task allocation if the RF front-end requirements can be individually specified for each of the separate blocks. For example, an LNA designer must meet all the requirements of noise figure, linearity, and gain with the restraint of input and output impedances being 50 ohms each. This scenario is illustrated in Fig. 4.4. However, a 50-ohm approach is not always optimal. In a receiver design, we are usually concerned about the voltage gain in the front-end, and interfacing the LNA with a high-input-impedance mixer can significantly boost the loaded voltage gain. This is shown in Fig. 4.5.

Placement of the filter before the LNA provides a better system design for rejecting out of band interferers and facilitates comparison of both the system architec-

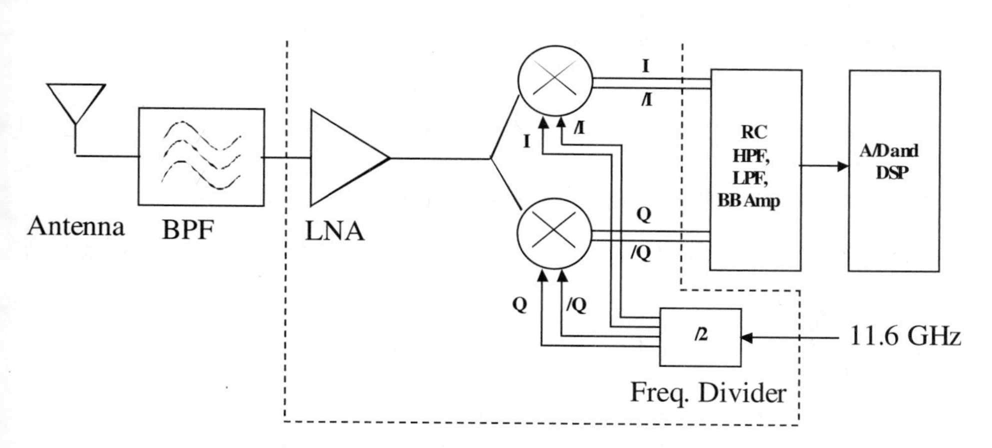


**Fig. 4.5.** Architecture II: Fully integrated receiver front-end architecture.

tures. However, placement of the filter in between the LNA and mixer helps to lower the noise figure of the entire receiver and reject out of band blockers. The dashed portions in Figs. 4.4 and 4.5 indicate the front-end circuit blocks that we will discuss. A differential configuration in the front-end mixers minimizes the effect of common-mode noise and provides the high IIP2 needed for direct conversion receivers. IEEE 802.11a modulation uses 52 information-carrying OFDM subchannels (each of which occupies 312.5 kHz) that are arranged symmetrically around a 53rd DC subchannel that contains no energy [4]. A padding of a 0.8 μs guard interval used for multipath protection makes the overall symbol period 4 μs [4].

Choice of a proper frequency scheme is important for a direct conversion receiver. Both the architectures under consideration here utilize a 2× frequency scheme, as shown in Figs. 4.4 and 4.5. In these cases, the incoming LO frequency is 11.6 GHz, twice the RF frequency (5.8 GHz).

### 4.1.4. System Calculations

The three key system specifications for RF front-end building blocks can be broadly categorized in terms of conversion gain (or loss), linearity, and noise figure. These specifications can be further fine-tuned into a set of specifications consisting of (a) conversion gain (or loss, (b) overall noise figure, (c) input P1dB, (d) IIP3, (e) IIP2, and (f) $I/Q$ imbalance, as appropriate. Also, the interfacing impedances play a big role in an integrated environment.

***4.1.4.1. Noise Figure.*** The noise figure referred to the antenna is calculated as follows:

$$NF = S - L - SNR - 10\log_{10}(BW) \tag{4.1}$$

where $S$ denotes the signal level; $L$ denotes the implementation margin to accommodate losses arising from channel estimation, $I/Q$ imbalance, filter distortion, and other nonidealities; $BW$ denotes the noise bandwidth (16.25 MHz); and $SNR$ is the signal-to-noise ratio given by

$$SNR = 10\log_{10}(E_b/N_0) + 10\log_{10}(E_s/E_b) \tag{4.2}$$

where $E_b$ and $E_s$ are the average energy per bit and symbol, respectively.

$$SNR = 10\log_{10}(E_b/N_0) + 10\log_{10}\{(\log_2 M) * R\} \tag{4.3}$$

where $R$ denotes the coding rate (1/2, 2/3, 3/4) and $M$ denotes the constellation density (2, 4, 16, or 64). Table 4.1 lists the SNR and noise figure (NF) requirements for various modulation schemes.

In the Table 4.1 $S_0$ is taken to be 8 dB more stringent than specified to take into consideration implementation losses. $E_b/N_0$ requirements for 10% BER are used for coherent demodulation.

**Table 4.1.** SNR and NF requirements for various modulation schemes in IEEE802.11a

| Data Rate (Mbps) | Modulation | Coding rate | $S_0$ (dBm) | $E_s/E_b$ (symbols/bit) | SNR (dB) | $E_b/N_0$ (dB) | NF (dB) |
|---|---|---|---|---|---|---|---|
| 6 | BPSK | 1/2 | −88 | 1/2 | 4 | 7 | 9 |
| 9 | BPSK | 3/4 | −87 | 3/4 | 5 | 6.25 | 9 |
| 12 | QPSK | 1/2 | −85 | 1 | 6.5 | 6.5 | 9.5 |
| 18 | QPSK | 3/4 | −83 | 3/2 | 8.5 | 6.74 | 9.5 |
| 24 | 16-QAM | 1/2 | −80 | 2 | 11 | 8 | 10 |
| 36 | 16-QAM | 3/4 | −76 | 3 | 14.5 | 9.73 | 10.5 |
| 48 | 64-QAM | 2/3 | −72 | 4 | 19 | 12.97 | 10 |
| 54 | 64-QAM | 3/4 | −71 | 9/2 | 20 | 13.46 | 10 |

***4.1.4.2. I/Q Imbalance.*** For the $I/Q$ imbalance calculations, the amplitude imbalance has been distributed in a geometrical mean fashion and the phase imbalance in arithmetic mean fashion between $I$ and $Q$ channels. Assuming an amplitude imbalance of $K$ and phase imbalance of $\theta$, the distorted version of a constellation with points $\{\pm a, \pm b\}$ is given by

$$I(t)_{bb} = a\sqrt{k}\cos(\theta/2) - b\sqrt{k}\sin(\theta/2) \tag{4.4}$$

$$Q(t)_{bb} = -(a/\sqrt{k})\sin(\theta/2) + (b/\sqrt{k})\cos(\theta/2) \tag{4.5}$$

SNR expression due to $I/Q$ imbalance is given by

$$SNR = \{\sqrt{k} + (1/\sqrt{k})\}^2 - 2\cos(\theta/2)\{\sqrt{k} + (1/\sqrt{k})\} \tag{4.6}$$

Since the constellations are all square and symmetrical, the C/N ratio remains the same for QPSK, 16-QAM, and 64-QAM. Hence, SNR for QPSK can be computed and the same C/N ratio expression can be used for other high-density constellations.

**Table 4.2.** $I/Q$ imbalance requirements for various modulation schemes in IEEE802.11a

| Data rate (Mbps) | Modulation scheme | Constellation error (dB) | Gain imbalance, as phase imbalance = 0 | Phase imbalance, as gain imbalance = 0 ($k$ = 1) |
|---|---|---|---|---|
| 12 | QPSK | −10 | 1.9 | 25.62 |
| 18 | QPSK | −13 | 1.36 | 18.14 |
| 24 | 16-QAM | −16 | 0.97 | 12.84 |
| 36 | 16-QAM | −19 | 0.69 | 9.09 |
| 48 | 64-QAM | −22 | 0.49 | 6.43 |
| 54 | 64-QAM | −25 | 0.34 | 4.56 |

The constellation errors are taken from the transmitter error vector magnitude (EVM) specifications. These specifications are derived from the high peak-to-average ratio of the OFDM signal. Often, the high instantaneous peaks in the time domain signal at the output of the transmitter are clipped by the PA. This is referred to as a hard-limiter, and is being adopted in some proposed solutions. Due to the signal clipping, the transmitted constellation gets distorted and the constellation error shows the extent to which such degradation can be tolerated.

To achieve adequate SNR for operation at 54 Mbps, an amplitude imbalance of 0.4 dB and phase imbalance of 3.7° provide a better design target.

***4.1.4.3. Linearity Requirements.*** Although IEEE 802.11a specifies a maximum power handling capability of –30 dBm (RMS) [4], a better design goal can be set at –25dBm (RMS). For a signal crest factor of 12 dB, the input P1dB should be more than –16 dBm (RMS). IIP3 is mainly related to the input compression point requirement for high input power signals and targeted IIP2 is set to exceed the minimum required value to enable operation at 54 Mbps also for high-power input signals.

A system target of a 10 dB noise figure with 5 dB of implementation margin has been set. $I/Q$ imbalance is targeted for 0.4 dB in amplitude and 3.7° in phase. Input P1dB has been set at –16 dBm. The overall front-end gain can be set at 20 dB as a compromise between the overall system noise figure and linearity.

The IEEE 802.11a wireless LAN modules operate in time division duplex (TDD) mode. Hence, there is no cross modulation involved between transmitter and receiver. This relaxes the LO–RF requirement on the direct conversion receiver. In the near future, a fully CMOS radio front end might be quite feasible to support a direct conversion receiver architecture. These modules should be targeted for very low power consumption receivers.

## 4.2. CIRCUIT DESIGN

All of the circuits have been developed with the SiGe BiCMOS process. The developed circuits blocks include an LNA, two different mixers, and a frequency divider. In this section, we discuss the circuit design steps in detail, with a brief introduction to the semiconductor technology in use.

### 4.2.1. SiGe BiCMOS Process Technology

The front-end design has been performed in a typical silicon germanium BiCMOS technology, which provides five metal layers with an additional thick metal layer for high-$Q$ inductor realization. The substrate resistivity is 10–20 Ω-cm and bipolar transistors with emitter widths of 0.32 μm, 0.44 μm, and 0.8 μm are available. The process utilizes high-performance graded bandgap-based NPN with $f_T$ and $f_{\max}$ of 47 and 60 GHz, respectively, for standard NPN BJTs, and $f_T$ of 27 GHz for high-breakdown NPN BJTs. It uses breakdown voltages of 3V and 5V, respectively, for standard and high-breakdown voltage devices. The high-$f_T$ transistors use a pedestal structure, which prevents spreading of the currents in different directions as opposed to the conventional BJT structure, thus improving the $f_T$. Gate oxide thicknesses of 5 nm and 7 nm used for MOSFETs (normal and 3.3 V, respectively). The NMOS and PMOS transistors have gate lengths of 0.24 μm. They support four to six layers of metal with the top metal being called the analog metal (AM), which is 4 μm thick and used for inductors and low-loss interconnects. Inductor $Q$ values are typically 19 at 10 GHz for 0.6 nH spiral inductors. Isolation mechanisms include deep trench (DT) and shallow trench (ST).

Having illustrated the process technology, we now focus on the different process features of this process in order to develop different front-end building blocks and integrate them.

***4.2.1.1. NPN Transistors.*** Normal NPN transistors in the process have beta = 100. The transistors can take three different width ($W$) values. Table 4.3 shows the different transistors available and lists considerations for their use in circuit design.

***4.2.1.2. Resistors.*** The two most commonly used resistors for front-end circuits are polysilicon and n-diffusion resistors. Table 4.4 shows the various parameters associated with these two types of resistors that must be taken into consideration when designing different parts of circuits. For example, when designing D flip-flops, the chosen load resistor is a polysilicon resistor to minimize parasitic capacitance. However, when designing current mirrors used for biasing, the n-diffusion resistors are chosen due to their lower tolerance limits.

### 4.2.2. LNA

For the LNA, a cascode topology is used with a gain switching option for bypassing a strong level of input signals at the receiver. The gain switch is implemented with the help of bypassing $n$FETs, whereas the high-gain cascode topology has been realized using bipolar transistors. The cascode topology has the inherent advantage of separating the output and input optimization criteria in the LNA circuit. Input and output matching are independent of each other. Also, it does not have the Miller capacitor effect, which is critical for high-frequency operation. Also, the reverse isolation is higher compared to the other topologies (cascade and single-ended topologies). The reverse isolation performance is very important for direct conversion

**Table 4.3.** Different transistor sizes and their impact on circuit performance

| Transistor width (μm) | Noise contribution from base spreading resistance | Miller capacitance | Process variation |
|---|---|---|---|
| 0.32 | lowest | highest | highest |
| 0.44 | medium | medium | medium |
| 0.8 | highest | lowest | lowest |

**Table 4.4.** Types of resistors and criteria for their choice

| Resistor type | Parasitic capacitance | Tolerance | Temperature coefficient |
|---|---|---|---|
| Polysilicon | less | more | ~ zero |
| N diffusion resistor | more | less | more |

receivers to keep the LO-to-RF leakage to a minimum. The minimum emitter width device has been chosen for the cascode transistors to keep the noise to a minimum value, as this lowers the base spreading resistance. The Miller capacitance increases with reduction of base width; however, the effect of Miller capacitor is mitigated in a cascode topology, as there is no possible coupling path from output to input. Use of inductive degeneration at the emitter leads to a wideband match at the input. Figure 4.6 shows the circuit schematic of the cascode LNA under consideration. Gain switching functionality has been realized by a NMOS transistor pair, M1 and M2. When the control voltage is "low" or 0 V, both M1 and M2 are turned off, and the LNA functionality is governed by the bipolar transistors Q1 and Q2. This is referred to as the "high-gain" mode operation of the LNA. The low-gain mode occurs when the control voltage is at a "high" state, typically raised to the supply voltage of the circuit. In this case, both the transistors, M1 and M2 are turned on. M2 provides a bypassing path to the input RF signal, and M1 provides a short between the input

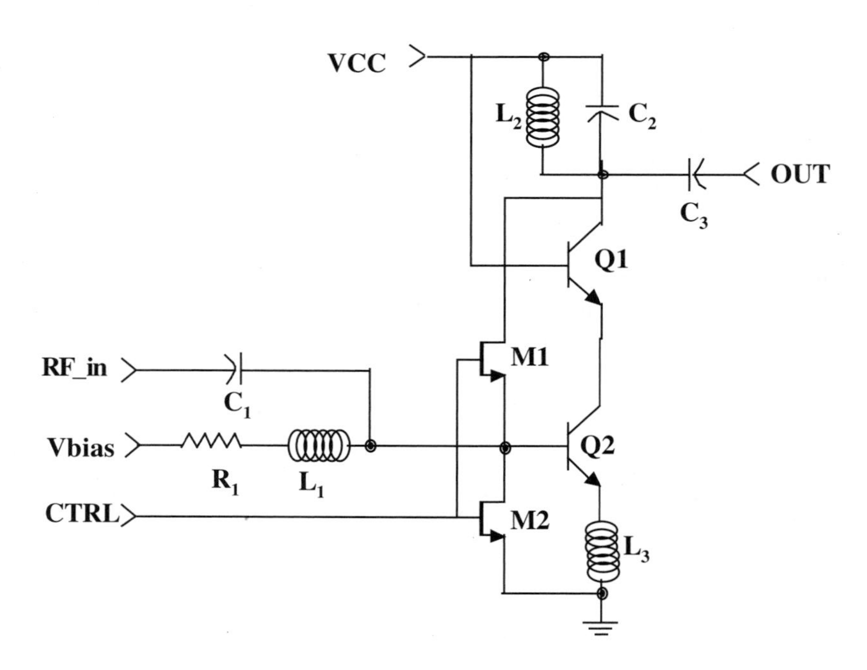


**Fig. 4.6.** Circuit schematic of a cascode LNA.

and output terminals. This ensures very little current flow to the base of the transistor, Q2, thereby reducing the gain significantly. This is called gain switching, by which, with respect to a control voltage, one can switch between two gain stages. One can also realize front-end amplifiers, which can operate at different gain stages with respect to control voltages; this is very useful for adjusting the signal strengths in the receiver front-end to avoid saturation of the circuits. Typically, the control voltages are determined by digital control from the baseband DSP, after the signal strength has been detected by the receiver signal strength indicator (RSSI).

For an input signal of a very high magnitude, the LNA gain is significantly lower than the high-gain operation, a necessity for wireless solutions to withstand high levels of interference. The input matching of the LNA is not required to be 50 Ω during the low-gain mode operation. The noise figure of the LNA is also not important in this case, as the main target is to bypass the signal, and not demodulate it or extract information from it. Gain switching is critical in the case of LNAs that are used for cellular applications, as in a mobile environment they can sometimes be closer to base stations. For a fixed environment (as in WLAN modules) gain switching is not very critical; however, it is good to provide gain switching capability.

Input and output matching is very important for the LNA, and it is very important to make the decision based on the system architecture and the impedances of the interfacing blocks. Usually, the input of the LNA is connected to the output of the band-select filter in the receiver front-end. Off-chip components are usually designed with an impedance of 50 Ω for both input and output. Thus, to ensure maximum power transfer from the filter to the LNA, one must make sure that the input of the LNA is matched to 50 Ω. The input matching consideration is the same for both the architectures under study. For the output matching, we have to consider the input impedance of the mixer. For a fully 50 Ω based receiver (architecture I), the output impedance should be 50 Ω. However, in the receiver, one should consider the voltage gain, and when a low-output impedance circuit is interfaced to a high-input impedance circuit, the voltage gain gets boosted significantly. This is the approach for architecture II—a non-50-Ω-based front-end.

A current source with current proportional to absolute temperature (PTAT) can also be implemented as a part of the LNA circuit. The gain of the LNA is governed by the transconductance ($g_m$), and denoted by $g_m = I_c/V_t$. $V_t$ is linearly proportional to absolute temperature. Hence, making the DC bias current proportional to absolute temperature helps keep the gain constant with respect to temperature variations.

### 4.2.3. Mixer

Two different mixer circuits have been utilized in the realization of direct conversion receivers. Both of the mixers are of Gilbert cell type. Both mixers utilize an active balun as a part of their RF transconductor section. They accept single-ended RF input from the LNA, and yield differential IF outputs in quadrature. This eliminates the need for separate baluns (on-chip or off-chip) in the receiver path. This is beneficial in terms of power consumption reduction and any noise figure degradation (due to the loss of passive baluns) in the receiver chain. The first mixer is a micromixer [7], which can provide a low input impedance stage and high linearity per unit of power consumption. This is used for architecture I, the 50-Ω-based approach. The second mixer is a single-ended Gilbert cell, which could provide high input impedance and can be utilized to boost the loaded voltage gain from the LNA to the mixer. A single-ended approach is quite compact, as no balun is necessary to generate differential signals. The balun circuitry usually introduces some amount of loss and imbalance in the RF signal path, which is critical for wideband operation in high data rate applications. The compromise is made in IIP2, which is lower in the case of a single-ended mixer, compared to its fully differented counterpart. For high-frequency applications, single-ended to differential conversion circuits in the mixer should be carefully optimized to minimize any imbalance.

The circuit schematic of the micromixer is shown in Fig. 4.7. The input stage of the micromixer consists of quasibalanced impedances containing a combination of inductor, resistor, and a transistor in a diode-like configuration.

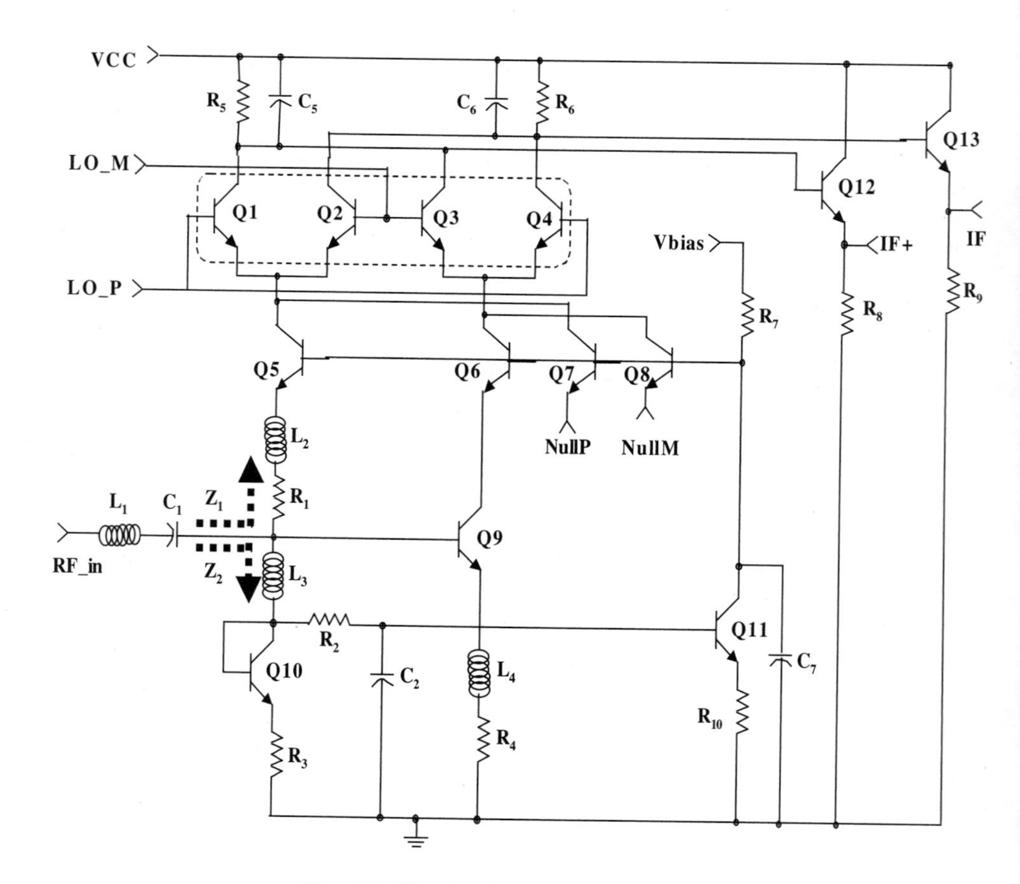


**Fig. 4.7.** Circuit schematic of a micromixer.

The input impedance of the circuit is formed by the parallel combination of the impedances seen in the two paths. These are formed by (a) the combined series impedance of L3, diode-connected Q10, and R3, and (b) the combined series impedance of R1 and L2, and the emitter resistance of Q5. Hence, the input impedance is given by

$$Z_{\text{in}} = (Z_1 \| Z_2) + j\omega L_1 + \frac{1}{j\omega C_1} \tag{4.7}$$

where

$$Z_1 = R_1 + \frac{1}{g_{\text{m5}}} + j\omega L_2 \tag{4.8}$$

$$Z_2 = R_3 + \frac{1}{g_{\text{m10}}} + j\omega L_3 \tag{4.9}$$

As seen from the above expressions, it is easy to realize a low-impedance input (50 $\Omega$) for the micromixer circuit. However, one should be careful with the balance between the two arms.

As two of the mixer inputs (*I*/*Q*) are connected in parallel, the individual input impedances should be set to 100 $\Omega$. Also, it utilizes a common-base (CB) stage formed by the Q5 and Q6 transistors, which provides high linearity per unit of bias current, as well as improved isolation between RF and LO ports. Q12 and Q13 form emitter-coupled buffers to isolate the mixer from the testing equipment. Capacitors C2 and C7 have been used as bypassing capacitors to minimize noise contributions from the bias circuit. Transistors Q7 and Q8, along with pull-down resistors, are used as offset nulling elements to tune out static DC-offset effects. The switching core is similar to a Gilbert cell configuration. The potential advantage of using a micromixer is its low power consumption and very high linearity performance.

Having introduced the micromixer-based circuit for architecture I, let us now focus on the single-ended Gilbert cell mixer for architecture II. Figure 4.8 shows the circuit schematic for single-ended Gilbert cell mixer. An inductively degenerated Gilbert cell with a bipolar differential pair at the input is used for each of the *I*/*Q* mixers of the fully integrated receiver, as the input stage usually exhibits higher impedance and, hence, is better suited for on-chip integration. One end of the differential pair has been grounded for conversion from single-end to differential operation. These active mixers operate at very low LO power, resulting in less possibility of LO to RF coupling due to the substrate and radiation.

Capacitors C2 and C4 have been used to minimize the noise contribution from the bias network. C3 has been chosen to be large to provide an AC ground to the other end of the Gilbert cell RF transconductor transistor pair. The input impedance of the Gilbert cell mixer (both of the *I*/*Q* mixers in parallel) can be combined with output impedance of the LNA and the signal amplitude should show a

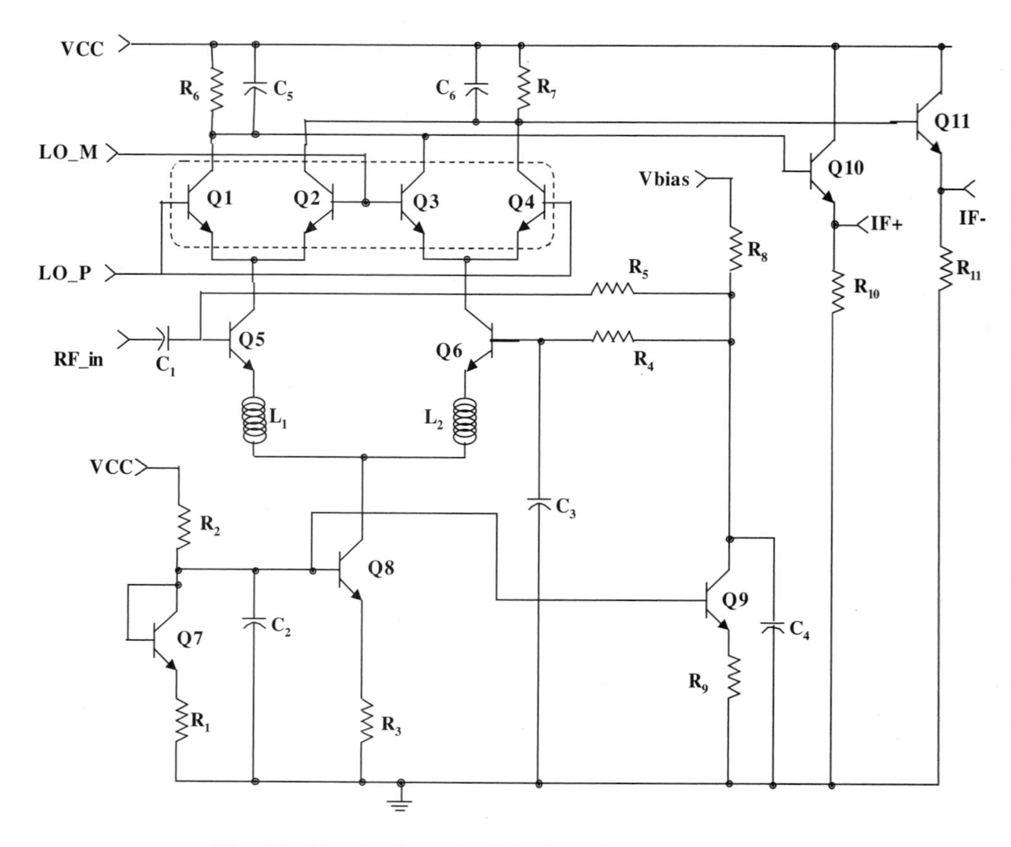


**Fig. 4.8.** Circuit schematic of single-ended Gilbert cell mixer.

peak at the LNA/mixer interface at the frequency of interest (in this case 5.8 GHz).

The MOSFET input stage provides high linearity compared to bipolars for a fully differential operation. However, MOSFETs have higher noise and thus degrade the circuit performance because, in a Gilbert cell, they are between the RF input terminals. Also, with the same current consumption, the gain obtained from MOSFET transistors is usually lower compared to that from bipolars and, thus, larger device sizes are required to provide the same gain. This larger-size again degrades the high-frequency performance of the devices.

Inductive degeneration helps achieve the linearity requirements while maintaining lower noise figures in the receiver. The operating frequency of the degeneration inductors should be less than (a) the self-resonating frequency (SRF) of the inductors (L1 and L2 in Fig. 4.8), (b) the resonating frequency of the network formed by the degeneration inductor (L1) and the base-to-emitter capacitance of the bipolar transistor (Q5) at the input stage.

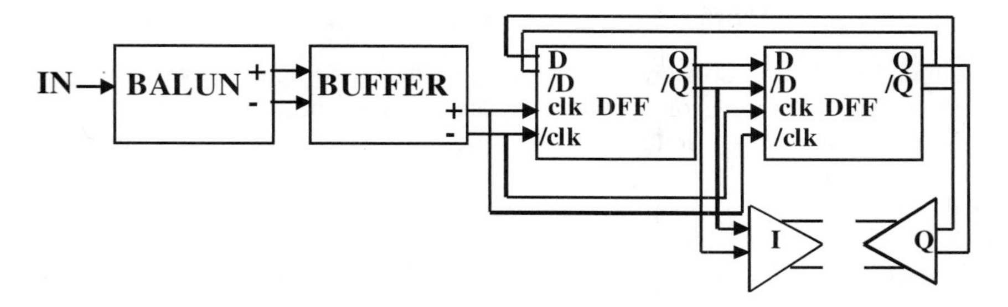


**Fig. 4.9.** Block diagram of a frequency divider.

### 4.2.4. Frequency Divider

Direct conversion receivers are very sensitive to LO-to-RF interference, which results in DC offset. Hence, the input LO frequency is at a different value (usually ½× or 2× the incoming RF frequency) to minimize LO-to-RF interference caused by the test fixtures and the probe cards to be used for testing. In the receiver designs under consideration, a 2× architecture based on a D flip-flop-based circuit has been used due to easy generation of quadrature signals (*I*, *I*/, *Q*, *Q*/). Another advantage of using a DFF-based frequency divider is accurate generation of quadrature signals. The scheme is shown in Fig. 4.9. It starts with generation of differential signals from a single-ended input, followed by the conversion to suitable voltage levels by appropriate buffers. The D flip-flop core generates quadrature signals, which in turn are amplified by the buffers to provide necessary LO swing to the mixer switching core. This architecture does not generate a frequency component at the subharmonic LO frequency, thereby improving LO-to-RF leakage performance. This circuit takes an input signal at 11.6 GHz and generates the four differential outputs in quadrature phases at 5.8 GHz. The input balun is designed on-chip and utilizes a single-to-differential conversion topology similar to the input stage of the micromixer shown before. The balun input impedance is a compromise between the input S-parameter and the power consumption, to realize the transconductance of the quasisymmetric input stage transistors, and targeted to a 50 Ω impedance match at 11.6 GHz, to interface with 50 Ω test equipment. However, careful optimization should be performed so that the balun generates accurate differential signals at 11.6 GHz.

## 4.3. RECEIVER DESIGN STEPS

This section provides a systematic illustration of the design of the receiver RF front-ends under consideration. We first introduce the various design steps to be followed in order. Then we introduce the use of simulators for a practical design and simulation of the receiver. Hence, the first part provides the reader with an overview of design and integration, and the second part provides an in-depth understanding of how to use computer simulators in the practical design process.

### 4.3.1. Design and Integration of Building Blocks

Different simulation tools are employed while designing RF receiver front-ends. The two broad categories of simulation tools are circuit simulators and electromagnetic (EM) simulators. Circuit simulators are used to simulate the active components (e.g., transistors) in conjunction with other passive components (e.g., resistors, capacitors) for designing front-end circuits. On the other hand, EM simulators are usually employed to accurately design and optimize passive structures for the front-end (e.g., inductors on silicon substrates, antennas, bandpass filters etc.). Inductors are critical in receiver circuits and often the standard silicon processes do not provide a scalable inductor model for IC design. Also, in a receiver design, it is necessary to use building blocks like passive baluns (flux-coupled type) for single-ended-to-differential conversion of the signal and vice versa. EM simulators help in the design, optimize, and provide a model circuit for the passive structures. The following sections describe the different features of the simulators and a sequence of simulations to follow to ensure the correct interpretation of the simulation results. For the receiver simulations, no EM simulation was used as the design kit was provided with a scalable inductor model.

There are two broad categories of simulators for IC simulation. The two most popular simulators in recent times are (a) Cadence and (b) Advanced Design Systems (ADS). Cadence is mostly utilized as a time domain simulator, whereas ADS is a frequency domain simulator. Also, as there is need for more and more integration on-chip, current silicon processes must have the capability to support digital, analog, and RF circuits. Thus, a simulation tool that is used to simulate circuits in a silicon-based process must be able to support codesign with highly dense digital (e.g., microprocessors etc.) and mixed-signal circuits. Sometimes, a mix of both simulators might prove optimal with regard to the designer's needs and available optimization time. For the direct conversion receivers presented in this chapter, all the simulations have been carried out with the Cadence simulation tool suite. The following section describes the sequence for simulation, which was followed to ensure a systematic design for the front-end.

### 4.3.2. DC Conditions

A DC simulation must be carried out prior to any other simulation to ensure proper bias points of the circuits. This ensures the operating points of the transistors in the circuit (cut-off, active, and saturated regions for the transistors), and also gives an idea of the power dissipation of the whole circuit. However, the bias points should be modified, depending on whether the circuit meets all the specifications regarding gain, noise figure, and linearity.

### 4.3.3. Scattering Parameters

To evaluate the maximum power transfer in the receiver front-end, scattering parameter (S-parameter) simulation with respect to a standard source impedance could be performed. If the architecture is based on building blocks of 50 Ω interfacing impedance (e.g., all the RF blocks provide 50 Ω for both input and output impedances), then S-parameter simulation must be performed by 50 Ω ports connected at the input and output of the circuits concerned. However, S-parameter is a small-signal simulation, and must not be employed in circuits where frequency conversion with respect to a large signal is performed (i.e., mixers). Usually, S-parameter simulation shows how closely the circuit's input or output impedance matches with a given port impedance. To ensure maximum power transfer from the test equipment to the circuit under test, the circuit's input impedance must be equal to the output impedance of the test equipment, which is typically 50 Ω. The amount of input match shows how much energy is reflected in interfacing the test equipment with the circuit (e.g., a 10 dB match indicates a 90% power transfer from the test equipment to the circuit, a 20 dB match indicates a 99% power transfer from the test equipment to the circuit). In the case of an integrated receiver simulation, S-parameter simulation should only be performed at the LNA input. 15 dB of input matching could be set as a design target for receiver inputs (too much optimization of input matching causes the input impedance to deviate from optimum noise matching, thus degrading the noise figure).

An S-parameter simulation with the inclusion of small-signal noise is quite suitable for a standalone LNA simulation. This simulation suite is sufficient for interpretation of the scattering parameters (S11 being the input match, S21 being the forward gain, S12 being the reverse isolation, S22 being the output match), in conjunction with the noise figure and the stability factor for the circuit.

### 4.3.4. Small-Signal Performance

Small-signal performance of the receiver front-end can be evaluated to ensure that the gain peaking occurs at the LNA/mixer interface, at the RF input frequency. If this gain peaking is not ensured, then the receiver will not be able to provide the optimum gain with the amount of power dissipation obtained by DC simulation. If the front-end system is a fully 50 Ω system, and both LNA output and mixer input are well matched to 50 Ω at the RF frequency of interest (a magnitude of 15 dB or more in the S-parameter), the gain peaking is automatically ensured at the LNA/mixer interface. However, for an integrated receiver, based on a non-50-Ω based interface, it is very important to observe the gain peaking. The transient simulation must be performed after the AC simulation, after one observes the gain peaking at the receiver front-end. An illustration is shown in Fig. 4.10.

The solid curve at the LNA/mixer interface is the desired frequency response. The dashed curve is the result of the AC simulation when the LNA/mixer interface is not tuned at the input RF frequency, Frf. This situation becomes worse for narrowband tuning, where the frequency response exhibits a sharp peak at a frequency other than Frf, and falls rapidly in frequency response. As is evident from the figure, if one performs a transient simulation prior to doing an AC simulation, the voltage gain might appear to be small, and this might give rise to a false indication that the current consumption of the circuit needs to be increased for higher voltage gain,

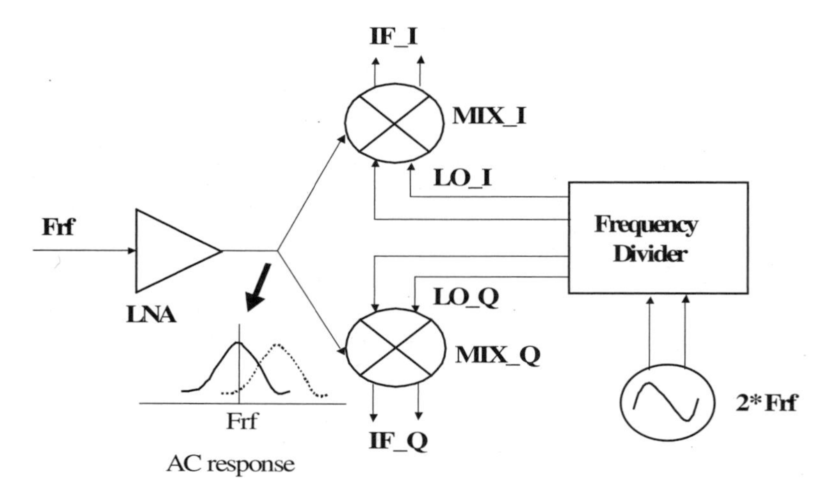


**Fig. 4.10.** Illustration of gain peaking at the LNA/mixer interface.

whereas it is actually the tuning at the LNA/mixer interface that needs to be optimized to obtain the maximum available gain for the given amount of power consumption. This is the reason why a small-signal AC simulation must be performed before doing a transient simulation.

### 4.3.5. Transient Performance

Transient simulation should be carried out to find out the voltage gain of the entire receiver front-end, in an integrated manner. This is especially useful to observe the effects of any harmonics resulting from the front-end, which are not captured by the small-signal evaluation, as described above. As the receiver front-end provides differential outputs, the voltage gain should also be observed by taking the difference between the output signals, both in-phase and quadrature. An $I/Q$ imbalance can also be observed by transient simulation. However, in the simulation we do not usually observe $I/Q$ imbalance. The source of $I/Q$ imbalance is layout asymmetry and any mismatch in amplitude and phase from the circuits at very high frequency.

### 4.3.6. Noise Performance

Noise performance of the receiver front-end is often evaluated with periodic noise simulation in terms of its noise figure. We provide an illustration of noise figure simulation in the voltage domain. Let us assume that at this point the receiver input has a reasonable amount of input match, so that its input impedance can be taken as 50 Ω. For a direct conversion application, the output noise spectral density should be integrated over the entire bandwidth of interest, and then averaged out. For a low-IF receiver front-end, spot noise (output spectral density at a particular frequency offset) should be observed. When the circuit is fully noiseless, the input source resistor (50 Ω) is the contributor of noise and gives rise to a voltage spectral density of 0.9 nV/$\sqrt{Hz}$. This is divided equally, and at the input of the circuit we would have 0.45 nV/$\sqrt{Hz}$ (by equal division). After the noise simulation, the input-referred noise can be calculated by transfering the output-referred noise to the input by dividing it by the voltage gain. The input-referred noise can be formulated as $V_{n,\text{in}} = V_{n,\text{out}}/A_v$. The noise figure can be formulated as: $NF = 20\log(V_{n,\text{in}}/0.45)$. The noise figure obtained by such an equation gives a double-sideband (DSB) noise figure estimate for the receiver front-end. Also, while integrating the output noise spectral density, one must integrate it over the required message bandwidth of the channel under consideration. In the frequency plan under consideration, the integration should be carried out from 156.25 kHz to 10 MHz in both sides of the zero frequency.

### 4.3.7. Linearity Performance

Linearity of the receiver front-end is often evaluated with the help of swept periodic steady-state (SPSS) simulation. This is the most time-consuming simulation, and it is recommended that one perform this at the end of the simulation sequence. Along with time, it requires a lot of simulation memory as well. This simulation can be carried out in both voltage and power domains. It is recommended that one provide a good S-parameter match (at the receiver input) before doing this simulation in the power domain (it can be performed without having a match as well, but with matching one can get linearity values very easily, using the built-in software macros in the simulator). The easiest way to visualize an integrated receiver is as an amplifier followed by a frequency downshifting in the spectrum. This observation is valid as long as there are some amplification and mixing operations, irrespective of their nature, such as actives and passives. The amplifier consists of the LNA and mixer combined gain, and the mixer switching core performs the downshifting function. In frequency domain, it is easy to imagine the IM3 tree formation, and then subtract the LO frequency in the frequency domain to yield an IM3 tree at the baseband.

All the simulation of linearity must be performed at the linear region of the receiver. A typical power level is –40 dBm for these simulations, as it falls well in the dynamic range of the receivers (which usually starts at ~ –100 dBm and goes up to –20 dBm). Three different linearity simulations are important: IIP3 (input third-order intercept point), input P1dB (1 dB compression point), IIP2 (input second-order intercept point). All of these parameters should be obtained separately in the simulation. Several rules of thumb to cross-check the results of the simulation follow.

In the linear region of the circuit, the fundamental and third-order intermodulation (IM3) tones at the output usually follow a slope of 1 and 3, respectively, when plotted in logarithmic scale with respect to the input power. However, the 3× slope of the IM3 products with respect to the fundamental is not true for all circuits, and one must be very careful when interpreting the results. It is important that one plot both the axes in some sort of logarithmic form (e.g., power in dBm unit for the x axis and voltage in dBV unit for he y axis) and extrapolate both curves to get the input third-order intercept point. Another rule of thumb, which is widely used, is given by the following equation: $IIP3 = P_{1dB} + 10$. This equation is also not valid in a lot of cases [8], so all the linearity parameters should be evaluated independently of each other.

The results and accuracy of the SPSS simulation depend significantly on the accuracy settings of the simulator and the set of frequencies utilized in the simulation. Cadence provides one with the flexibility to choose voltage, current, and absolute tolerance values in a simulation. The tighter the tolerance values, the more accurate the simulation results become. However, a stringent set of tolerance values leads to an increase of simulation time and memory for such simulations. A trade-off must be achieved earlier in the simulation process. First, one should analyze the circuit theoretically to ensure whether the fundamental and IM3 would follow slopes of 1 and 3, respectively, with respect to input power. Then some quick simulations should be carried out and the linear region of the receiver should be detected. When this is done, one should think of tightening the tolerance parameters to simulate and obtain fairly accurate results of IIP3 and P1dB. It can be helpful at this point to cross-check the 10 dB difference equation between P1dB and IIP3.

The second-order input intercept point (IIP2) is very critical for direct conversion receivers. This is usually obtained by extrapolating both the fundamental and second-order intermodulation (IM2) components; the input power level at the intercept point is denoted by IIP2. However, IIP2 is a function of device mismatch, and should be simulated with some amount of device mismatch introduced in the mixer transistors. A 3% mismatch between devices in a semiconductor process is quite typical, and could be taken as a starting point for IIP2 simulation. Without the presence of device mismatch, the second-order (even-order) terms are cancelled differentially at the output, and any interpretation of IM2 would be faulty (in this case, the simulation would show an extremely high value of IIP2, which is theoretically not justified). As one introduces increasingly higher mismatches in the devices, the IIP2 gets worse. Also, the IM2 can only be generated from the front-end mixer, as any IM2 product generated by the LNA is heavily attenuated (or blocked) by the coupling capacitor at the LNA/mixer interface.

The choice of frequencies for SPSS simulation is also very important, as an improper choice can either require significantly higher simulation time, or produce incorrect results. The cadence simulation tool suite first finds the beat frequency from the different sets of frequencies used in the simulation tool suite (two closely spaced RF frequencies and one LO frequency). Then it tries to find periodicity in the various waveforms during the simulation, until the steady state is achieved. If the beat frequency is lower, the simulator takes more time in converging toward its final solution. However, if the beat frequency is higher, there is the possibility of the IM3 tree falling onto the IM2 product. As an example, let us assume that the two RF frequencies of 5.83 GHz and 5.84 GHz were input to the receiver operating with LO frequency of 5.8 GHz. In this case, the IM3 tree at the output would consist of (20 M, 30 M, 40 M, 50 M). However, IM5 terms would be present at the frequencies of 10 M and 60 M, respectively. At the same time, the IM2 term is also at frequency of 10 M (5.84 G–5.83 G). Hence, with this set of frequencies, the IIP2 simulation results would be incorrect. However, if we had chosen frequencies of 5.803 G and 5.804 G, then there is no chance of inaccuracy due to one of the IM components falling on a IM2 component, but it would require a significant amount of simulation time. This situation is illustrated in Fig. 4.11.

When the frequencies used for two-tone test are far apart, there is a possibility that the upper part of the IM tree might get attenuated due to the low-pass frequency response of the RC load present at the mixer output. Plotting the fundamental component from the upper side of the IM3 tree (which could be attenuated by the low-pass nature of the mixer output), and the IM3 component from the lower side of the tree (not attenuated by the RC low-pass filter) would result in a poor value of IIP3, whereas choosing the fundamental component from the lower side of the IM3 tree (not attenuated) and the IM3 component from the upper side of the IM3 tree (attenuated) would result in a higher value of IIP3. Thus, while plotting the fundamental and IM3 components, one should plot these components either from the upper side or the lower side of the IM3 tree in the down-converted spectrum, to incorporate the effects of attenuation in a similar way for both the fundamental and IM3 components. Usually, the components in the lower side of the IM3 tree could be used to ensure least possible attenuation from the filtering action.

It must be noted that when using ADS for simulations, harmonic balance simulation is used to produce the results that SPSS and periodic noise simulations generate in Cadence.

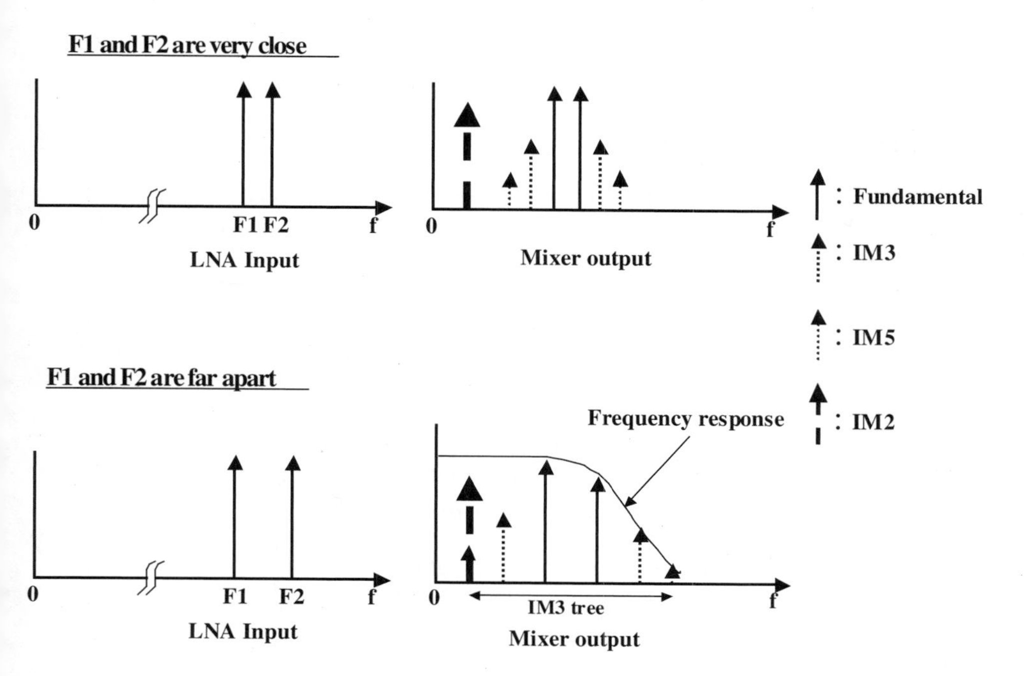


**Fig. 4.11.** Illustration of frequency choice for SPSS simulation.

### 4.3.8. Parasitic Effects

At high frequencies, the parasitics become extremely important to consider. After performing the layout of the entire receiver front-end, it is necessary to find out the parasitic capacitances and resistances at various circuit node points and interconnect traces. At extremely high frequencies (of the order of tens of gigahertz), these parasitics can be utilized for matching purposes as well. Parasitic elements usually consist of capacitors, resistors, and inductors. However, the latter two (resistors and inductors) are very difficult to extract in a three-dimensional geometry, and time-consuming. Hence, for the extracted simulation, only parasitic capacitance could be considered to save time. Simulations performed with extracted parasitics deviate from the circuit simulation with regard to the matching, noise figure, and gain. However, the linearity performances remain the same for both of the simulations (extracted and circuit schematic netlists). A detailed discussion on layout parasitics is presented in Chapter 7.

### 4.3.9. Process Variation

The model parameters (e.g., threshold voltages of MOS transistors, resistor values, etc.) of the circuit may vary to some extent while in fabrication. This situation gives rise to mostly three different model parameters: slow, typical, and fast. These are usually termed process corners, and it must be ensured that the all the simulations pass the process corners with a reasonable margin. In the receiver front-end under consideration, the most critical block is the frequency divider, and, thus, it is mandatory to ensure that it passes all the corner simulations.

### 4.3.10. 50-Ω and Non-50-Ω Receivers

The above steps provide a detailed and systematic design procedure for any integrated circuit/receiver design. For a 50 Ω-based architecture in the front-end, the above steps can be applied on each of the building blocks. The building blocks in architecture I—for example, LNA and micromixer—have been developed by applying the above design steps on the respective circuits. For a non-50 Ω-based receiver, as in architecture II, the circuits, which are non-50 Ω-based interfaces, should be simulated in the voltage domain (and not in the power domain) and optimized for performance. When the designer reaches an optimum point, the circuit should then be integrated with the building blocks, which are 50 Ω based interfaces. For example, in case of an LNA/mixer integration, the LNA input should be designed as a compromise between noise and a 50 Ω match. The mixer should first be designed with simulation methodologies in the voltage domain. After an initial optimization, the mixer should be integrated with the LNA and the interface should be carefully designed so that the loaded voltage gain peaks at the interface. The other design methodologies and optimization steps are very similar to the ones described above.

## 4.4. LAYOUT CONSIDERATIONS

For any high-frequency design, the layout plays a very important role in the circuit performance. This is more critical in the cases of silicon-based processes, due to their lossy nature. There are many generic design guidelines for mixed signal layout, but in this section we will consider only those that are pertinent to the chosen application. Since the circuit consists of both analog and digital building blocks, care must me taken to separate the two power supplies, thus minimizing the effect of any common-mode noise from the power supplies. Also, the digital circuitry (frequency divider in this case) should be surrounded by deep-trench and substrate contacts to minimize the noise generated by digital switching on the other analog circuit blocks of the receiver. Substrate contacts (also referred to as "taps") also help in isolating one circuit block from another by fully absorbing the ground current in the substrate.

The input signals of the receiver are taken in the topmost metal available in the technology. The input lines from the pad are kept short to minimize the parasitic capacitance resulting from the layout traces. The LNA noise figure is very sensitive to the deviation in input impedance and, hence, if the layout traces contribute much parasitic capacitance, the noise figure will degrade.

As per the inductors, any two inductors used in the circuit should be placed far apart from each other to minimize any flux coupling effect. A separation of 60 μm has been adopted for the designs and has been followed in the LNA and mixer layouts. The high-frequency interconnect lengths should be kept as small as possible and should be preferably done in AM or MT (top metal) to reduce resistance, capacitance, and inductance. This holds for the interconnections between various cells (like LNA to mixer).

Shielding of the input signals from the substrate is also very important. This is done by having the input signals in AM and providing the ground as stacked up M1 and M2 running as a big thick ground surrounding the chip and underneath the pads.

Layout symmetry is extremely critical in receiver front-ends. *I*/*Q* imbalance associated with the mixers is very sensitive to the layout symmetry. Hence, in the layout, equal trace length must be assured from the RF input to the mixers. This is even more critical in the D flip-flop layout in terms of the clock distribution network. If the transistors are not identical and the layout traces are not equal, it results in deviation of the duty cycle from 50%.

In high-frequency operation, parasitic capacitances must remain at a minimum level. To ensure this, a deep-trench (DT) maze on the polysilicon (PC) resistors should be used. This is followed in the frequency divider layout where the input frequency is 11.6 GHz and any extra parasitics degrade the performance significantly. The DFFs in the frequency divider utilize collector resistances built on a DT maze. Also, the resistor at the input of the LNA needs to be built on a DT maze to reduce the additional parasitic capacitances.

Sometimes, shielding of the layout traces becomes very important. If the metal line connecting the base goes underneath the metal line connecting the collector and vice versa, it gives rise to additional Miller capacitance, which might degrade the high-frequency performance. Hence, another metal layer can be used as shielding for the two traces.

Large DC blocking capacitors are provided between the power supply and ground to minimize noise perturbation from the supply voltages as well as to prevent any kind of ringing effect from DC supplies. The building blocks for both of the receiver architectures are laid out for the on-wafer probing technique.

## 4.5. CHARACTERIZATION OF RECEIVER FRONT-ENDS

In this section, we provide the reader with a systematic sequence for the measurement of the developed receiver front-ends. The measurement setups were performed in exactly the same way the front-end was simulated. The LNA is characterized by S-parameters, linearity (P1dB and IIP3), and noise figure (NF) parameters, whereas both the mixer and integrated receiver are characterized for conversion gain, linearity (P1dB, IIP2 and IIP3), *I*/*Q* imbalance, 50-ohm-input match, noise figure, and DC offset. A flow chart for complete receiver testing is demonstrated in Fig. 4.12.

The fabricated ICs are characterized using the on-wafer characterization technique. Two types of testing methods have been used for measurements: (a) probe cards, and (b) wedge probes. Probe cards are custom made with a number of power supply signals and RF signal ports, and have been used for testing the micromixer circuit as described above. However, probe cards are custom made and the pad

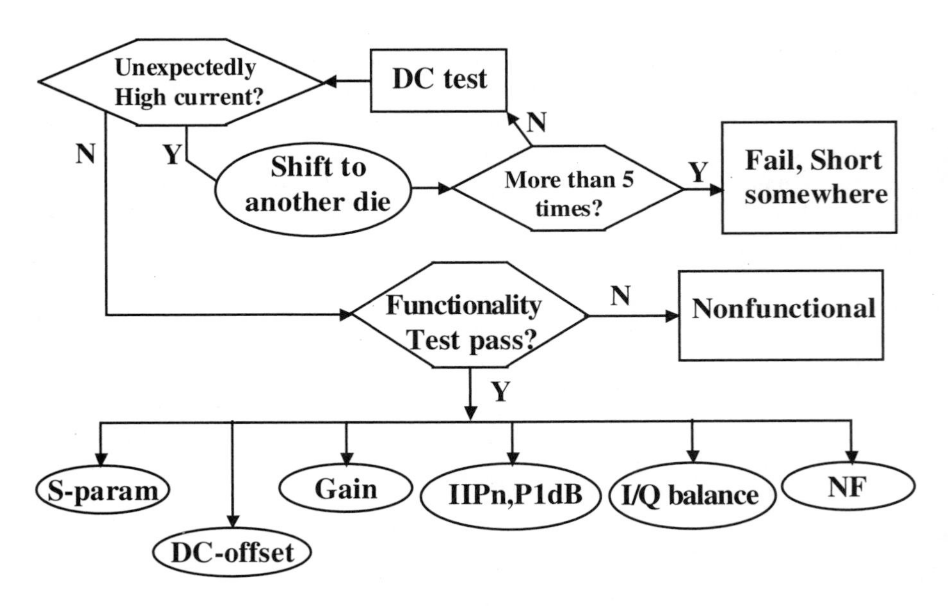


**Fig. 4.12.** A complete flow chart for mixer/receiver testing.

frame must be laid out accordingly. Wedge probes offers flexibility in terms of their positioning, and four such probes have been used for testing the integrated receiver.

### 4.5.1. DC Test

The first step to characterize any circuit is to set up the DC bias points for the circuit under test (labeled as receiver under test or RUT in the following figures), and observe the DC currents flowing through the IC. The bias current can be monitored by using a multimeter as an ammeter. The supply voltage in the circuit should be adjusted to the nominal value as in simulation. If the current through the circuit is much higher than expected, there might be a short in the circuit, and the DC operating condition will not be satisfied. In this case, the probe card should be shifted to another IC sample. However, if the circuit passes this test, then other performance tests should be carried out. Also, when the circuit is functional, the DC test arrangement can be used for splitting up the current consumption for different receiver blocks. For example, a DC test on the entire receiver would indicate the power dissipation of the LNA, mixer, and frequency divider separately.

### 4.5.2. Functionality Test

After the DC bias points have been set up, a test should also be carried out to determine whether the circuit is functional. For a mixer, it must down-convert the incoming RF signal to a desired IF frequency. The bias values can be adjusted a little bit to obtain the proper functionality of the circuit. If the circuit does not pass this test, no further tests should be carried out. In this specific test, two frequency sources, RF and LO, are used at frequencies of 5.8001 GHz and 11.6 GHz (an on-chip divider generates 5.8 GHz signal) and an oscilloscope is used to observe the 1 MHz down-converted IF signal. This provides the indication that the circuit is functional as a down-converter. Other tests can be performed to characterize both the down-converter and the mixer.

### 4.5.3. S-Parameter Test

A one-port S-parameter test is performed on both the receiver and the down-converter to observe the matching at the input port of both the circuit blocks. The calibration for this measurement is performed off-chip, and the receiver under test (RUT) was tested. This test should be performed prior to any other tests in the receiver, to make sure of the fact that maximum power transfer occurs from the RF signal generator to the RUT. As shown in Fig. 4.13, only DC supplies are required for the one-port S-parameter test.

### 4.5.4. Conversion Gain Test

The test bed for receiver gain testing is shown in Fig. 4.14. For convenience of testing, an IF frequency of 1 MHz is used. Off-chip accessories, such as differential-to-

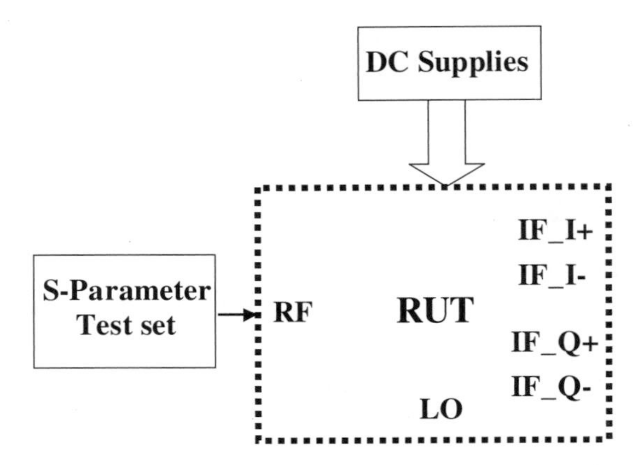


**Fig. 4.13.** Test setup for one-port S-parameter.

single-ended converters are used to interface with the testing equipment. The on-chip differential buffers drive the inputs of the OP-amp based circuitry, which is a high-impedance stage. The output impedance of this OP-amp stage is 50 Ω, which interfaces with the spectrum analyzer. Gain can be calculated as the difference between the spectrum analyzer reading to the amplitude of input RF signal. Tests on both *I* and *Q* channels should be done to cross check the validity of the measured conversion gain across the channels. Differential-to-single-ended converters usually exhibit high linearity, so that the RUT does not saturate them.

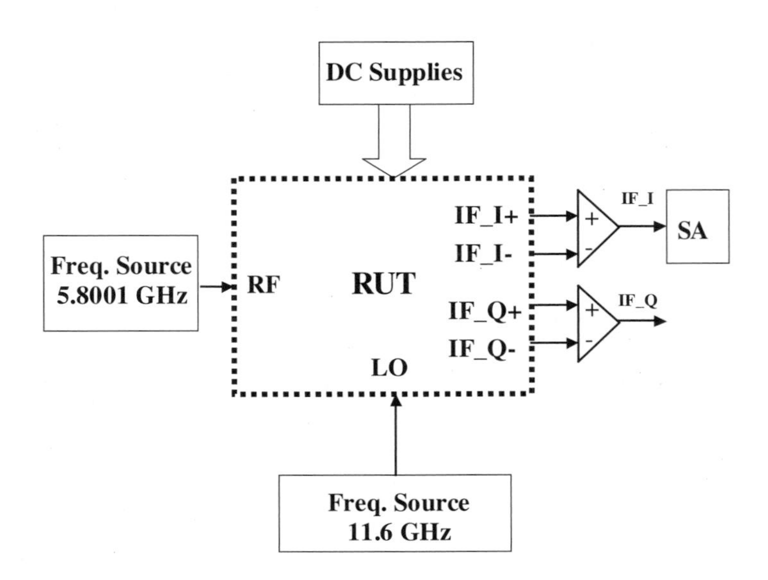


**Fig. 4.14.** Test setup for receiver gain measurement.

*Note.* During the testing process, the gain from the additional off-chip OP-amp circuit should be de-embedded to measure the entire circuit gain. The one used in this test had a –6 dB gain.

### 4.5.5. Linearity Test

The linearity test setup is shown in Fig. 4.15. Two closely spaced tones at 5.8005 GHz and 5.8006 GHz were combined off-chip and passed through a variable attenuator for signal strength control to the RUT. The outputs corresponding to second- and third-order distortion terms (IM2 and IM3) are seen in the spectrum analyzer. For the circuits under consideration, the IM3 should follow a slope of 3 with respect to input power, and the IM2 should follow a slope of 2 with respect to input power. The variable attenuator is used to control the level of input power going through the RUT. Once again, the off-chip OP-amp buffers must have more linearity to make sure that no additional distortions crop up due to their presence. Also, the frequency source for the LO should have a very low phase noise. The testing should be carried out until the IM5 terms appear at the output.

*Note.* In the usual case, the IM2 products should rise with a slope of 2 for each unit increase in the input power, and IM3 products should rise with a slope of 3 for each unit increase in the input power. Some circuits, however, do not behave this way. The circuits under consideration have slopes of 2 and 3 with each unit increase in input power, and this can be used to cross-check the measurement results. Also, IM2 is an even-order product resulting from device mismatches, and varies from one circuit to another.

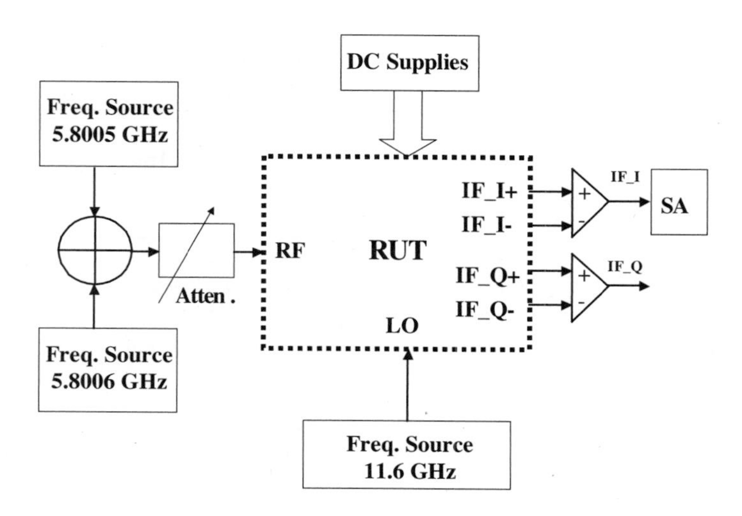


**Fig. 4.15.** Test setup for linearity measurement.

### 4.5.6. Noise Figure Test

The test setup is shown in Fig. 4.16. At the output of the differential IF terminals, a transfomer circuit (or balun) is placed for the single-ended-to-differential converter. OP-amp-based differential-to-single-ended converters cannot be used in this case, as they have a very high noise contribution, which can swamp the noise contribution of the RUT. Transformer-based circuits do not provide any gain, but keep the noise contribution to a minimal value. The noise figure meter should be calibrated prior to testing. Also, the NF measured by the NF meter is a double sideband (DSB) noise figure, relevant for a direct conversion receiver. This noise figure can also be measured versus frequency by sweeping the RF frequency.

### 4.5.7. *I*/*Q* Imbalance

The test setup for *I*/*Q* measurement is shown in Fig. 4.17. Four identical cables from the RUT IF terminals (IF_I+, IF_I–, IF_Q+, IF_Q–) are used to interface with the oscilloscope. It is important to use exactly identical cables in this test, as the imbalances in the cables might otherwise introduce some *I*/*Q* imbalance during the measurement. The high-input impedance of the oscilloscope can be used for measurement. On the oscilloscope screen, it is possible to obtain the waveform statistics (amplitude and phase information, along with their mean and variance values). From this information, the *I*/*Q* imbalance can be extracted.

### 4.5.8. DC Offset

The test setup for DC-offset measurement is shown in Fig. 4.18. The positive and negative outputs from the RUT-IF terminals are interfaced with a multimeter to

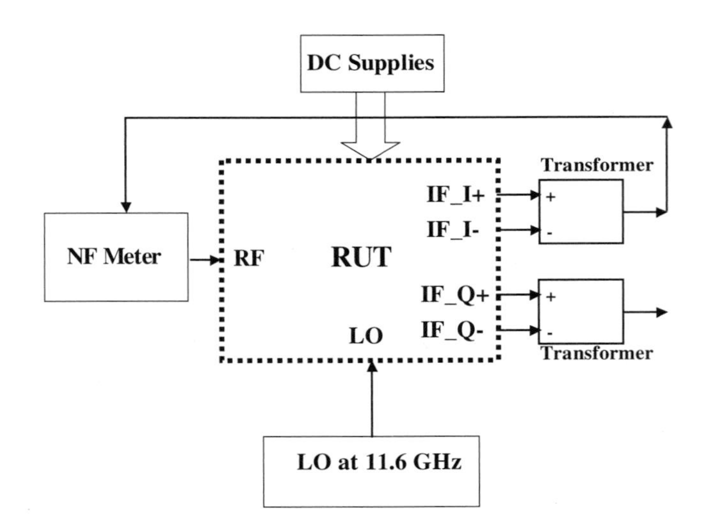


**Fig. 4.16.** Test setup for noise figure measurement.

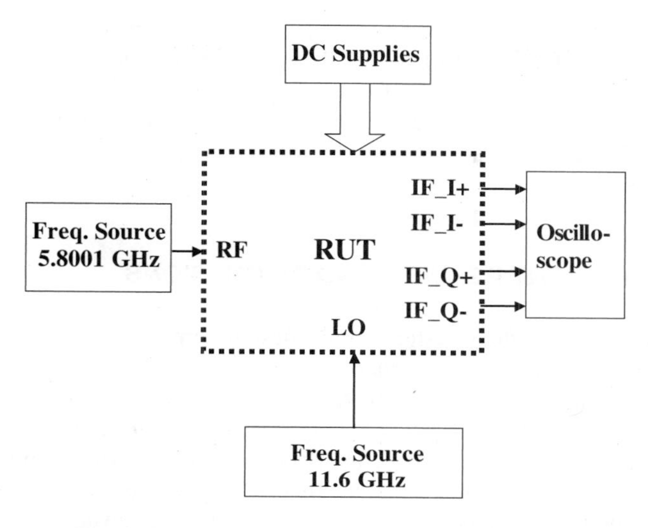


**Fig. 4.17.** Test setup for $I/Q$ imbalance measurement.

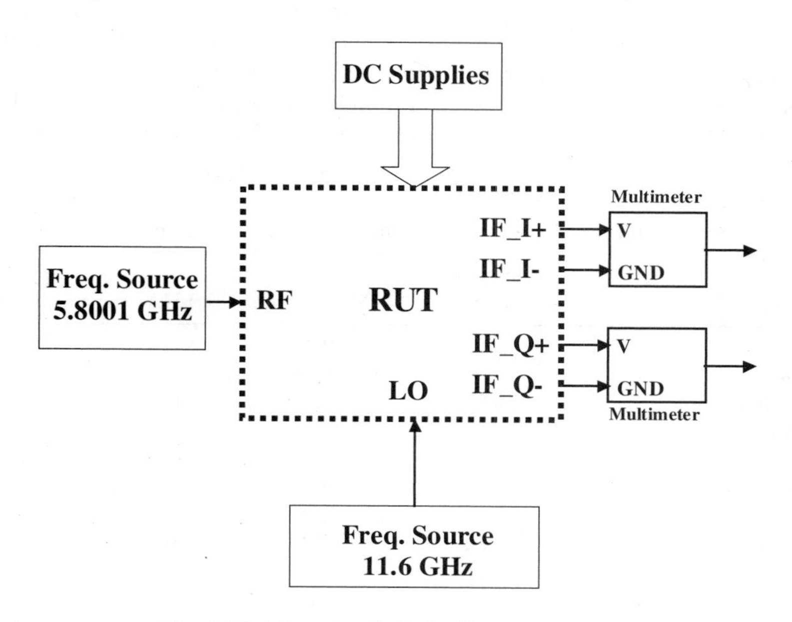


**Fig. 4.18.** Test setup for DC-offset measurement.

measure the voltage difference. The same test setup should be used to measure both static and dynamic DC offset of the RUT. For the static DC offset, LO and RF sources are turned off and the DC offset is measured. For the dynamic offset, either RF or LO or both should be turned on one by one to observe the dynamic DC-offset value. Static DC offset is also dependent on device mismatch, and varies from one chip to another.

## 4.6. MEASUREMENT RESULTS AND DISCUSSIONS

Having explored the system design, circuit design, simulation, and measurement techniques, we now provide the reader with the specifics of measurement results. We provide an explanation of deviations wherever possible. Performances of the ICs have been measured across different chips (usually five to six in number). Characteristics such as DC offset and IIP2 are dependent on mismatches present in the devices, and can vary from one chip to another. The following sections describe the different measurement results obtained from the various hardware blocks fabricated.

This section compares the performances of two different architectural options chosen for the IEEE 802.11a specification. A reasonable estimate of front-end passive losses should be incorporated in both cases to estimate the overall performance of both receiver architectures. One common advantage of both the architectures is that they do not require a separate on-chip balun, as they both utilize single-ended mixers and, hence, are very compact. Table 4.5 shows the performance summary of the fabricated ICs along with architectural trade-offs. Figures 4.19, 4.20, and 4.21 show the die photographs of the front-end subblocks. For architecture I, the overall

**Table 4.5.** Performance summary of developed receiver

| Performance | Integrated receiver | Micromixer | LNA | LNA + micromixer* |
|---|---|---|---|---|
| Gain (dB) | 20.2 | 9.23 | 11 | 20.23 |
| NF (dB) | 7.1 | 19.5 | 4.4 | 9.89 |
| ICP (dBm) | –15.5 | –3 | –10.4 | –14.0 |
| IIP3 (dBm) | –3 | +6 | –2 | –6.76 |
| IIP2 (dBm) | +31 | +32 | — | +21.0 |
| Amplitude imbalance (dB) | 0.9 | 0.25 | — | |
| Phase imbalance (deg) | 5 | 2.4 | — | |
| Power consumption (mW) | 64 | 32 | 16 | 48 |
| Input match (dB) | 15 | 16 | 20 | |
| LO power (dBm) | –10 | –12 | — | |
| LO-to-RF leakage | 78 dB | 58 dB | — | |
| Dynamic DC offset (mV) | <1 | <1.6 | — | |
| Output match (dB) | — | — | 12 | |

*Denotes computed results from individual chip results.

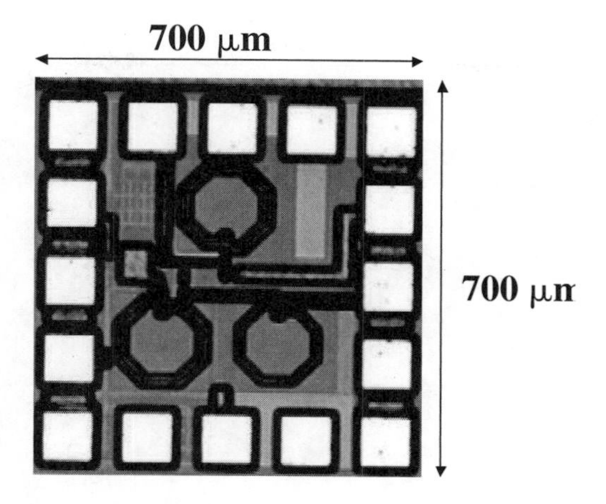


**Fig. 4.19.** Die photograph of LNA.

performance is computed from the measurement results of the LNA and micromixer. The conventional 50 Ω system (architecture I) provides a higher noise figure due to the high noise figure of the micromixer, as well as the LNA. However, the power dissipation is lower for the 50 Ω architecture, with similar linearity and conversion gain values. The developed ICs can be readily integrated with LTCC-based front-end passives [9] for entire front-end solutions. The LNA exhibited an upward frequency shift (operates at 6.4 GHz) and higher noise figure (4.4 dB) compared to its

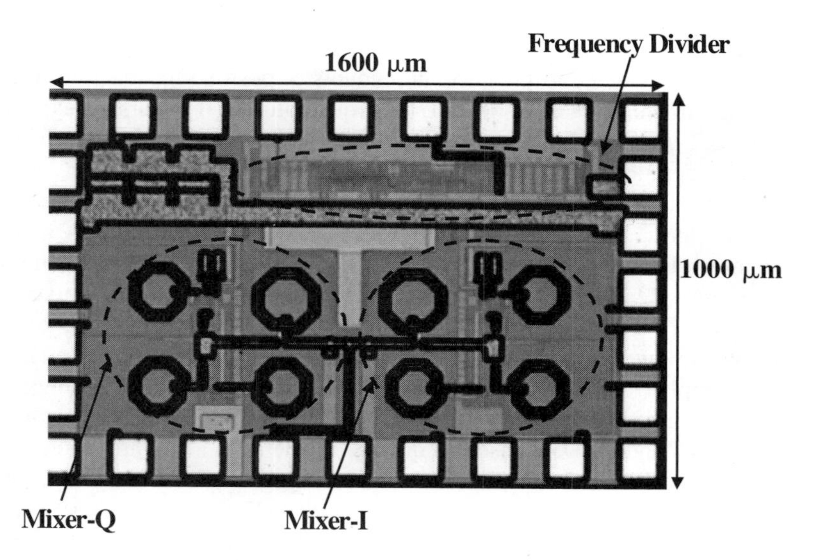


**Fig. 4.20.** Die photograph of fully monolithic micromixer IC.

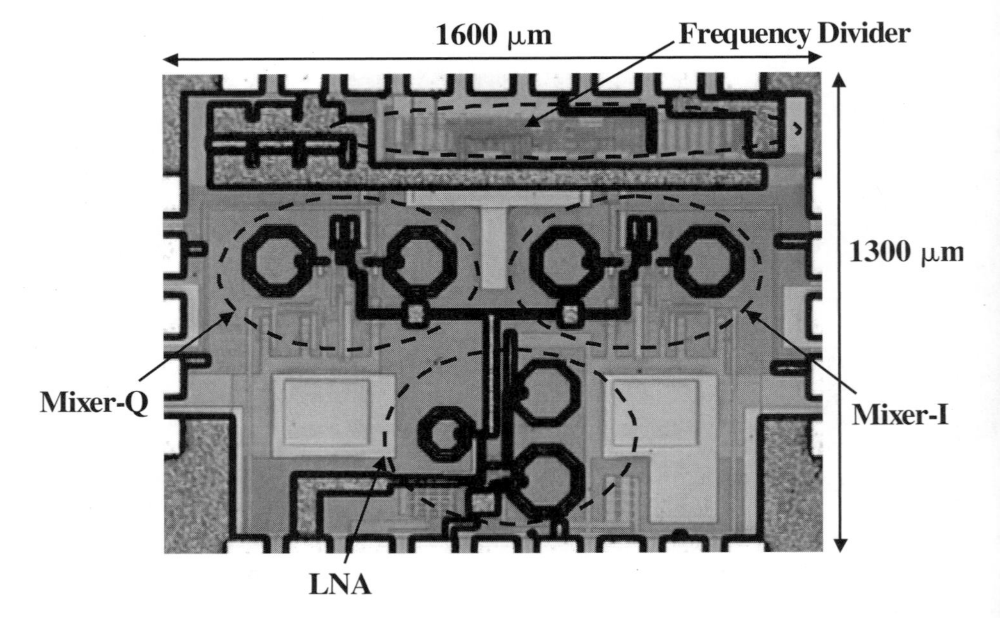


**Fig. 4.21.** Die photograph of fully monolithic receiver IC.

simulated performances. The reasons for these discrepancies have been explained in detail in Section 4.6.1. However, the LNA performs better in the fully integrated receiver IC. The developed receiver front-end for architecture II meets both the noise figure and linearity requirements with a 5 dB implementation margin. The *I*/*Q* imbalance is higher in measurement compared to expected performance from simulations, and is attributed to the unequal amounts of positive feedback from the on-chip ground in the cases of in-phase (*I*) and quadrature (*Q*) mixers, as explained in Section 4.6.2. This sets the stage for an interested reader to wonder how critical the layout considerations can become at such high frequencies. However, amplitude and phase balance specifications can be relaxed if the DSP compensates for their distortion effects. In the integrated receiver (architecture II), the LNA and mixer combination dissipates 64 mW of power and the frequency divider operating at 11.6 GHz consumes 22 mA of DC current from a 3.75 V supply. The micromixers (*I* and *Q*) dissipate 32 mW of power, and the LNA dissipates 16 mW of power, giving rise to 48 mW of power dissipation in LNA and mixer circuits (architecture I). Use of different supply voltages on different parts of the circuits is also a concern, and should be optimized further for 2.7 V. Also, 11.6 GHz is a fairly high input frequency for a standard DFF-based frequency divider and a ½× architecture (input frequency of 2.9 GHz) shows potential for improving LO to RF isolation as well as power dissipation and can be adopted for the targeted application. This would be a better choice compared to a 2× architecture due to the proposed DC-free modulation scheme of IEEE 802.11a.

### 4.6.1. Close Examination of Noise Figure and *I*/*Q* Imbalance

From the measurement results provided in the last section, the inquisitive reader might wonder about the high noise figure of the LNA and the high value of *I*/*Q* imbalance for the integrated receiver. Noise figure values are very different from what was simulated (2.7 dB). The following arguments can possibly explain this deviation.

(a) During the fabrication of the circuits, the inductor values were changed (reduced) by a 10% value. This caused the frequency peak to occur at 6.4 GHz rather than 5.8 GHz. This would cause a significant deviation in the frequency response, and cause the optimum impedance necessary for noise figure to shift. However, as the LNA is capable of providing large bandwidth, the gain values at the two frequencies—5.8 GHz and 6.4 GHz—only varies slightly (~ 0.4 dB).
(b) The input inductor must provide a high quality factor (*Q*). The *Q* value decreases when an inductor is confined by metal lines in the layout. Also, the metal lines should be put farther apart from the inductors to avoid any induced current flow.
(c) The metal lines used to provide DC bias could contribute some amount of inductance at RF frequencies if they are very thin.
(d) The extraction tool sometimes does not account for the resistances contributed by metal connections on the emitter and collector for the transistors. It only models the intrinsic resistance between polysilicon and base.
(e) It is desired to design inductors with a nonintegral number of turns (e.g., number of turns = integer + three-fourths or one-fourth). Input and output terminals of the inductors should be taken at different positions to minimize the effect of input-to-output coupling. In the receiver design, the input inductors were taken to be of an integral number of turns.
(f) The noise models used for the simulation may not be very accurate at 5.8 GHz.
(g) Usually the probe cards account for 0.1 dB of noise figure. The calibration mechanism was done off-chip and, thus, the effect of probe card (~ 0.1 dB) should be de-embedded.

To account for all of the above steps, an off-chip inductor can be used. This is a critical inductor in the receiver which would require a very high *Q*, and an off-chip would be a much better option than an on-chip inductor. Also, some other matching technique, in which the input inductor is not needed, might be desired as well.

### 4.6.2. Comments on *I*/*Q* Imbalance

The measurement result of *I*/*Q* imbalance was unexpectedly higher, and was found to be consistent among different chips that were tested. Since the layout had been done symmetrically, there could be two possible explanations for this effect.

One of the reasons for *I*/*Q* imbalance is the imbalance in the quadrature signals generated from the frequency divider. As explained earlier in Section 4.2.4, the input stage (balun) for conversion from single-ended to differential signals is

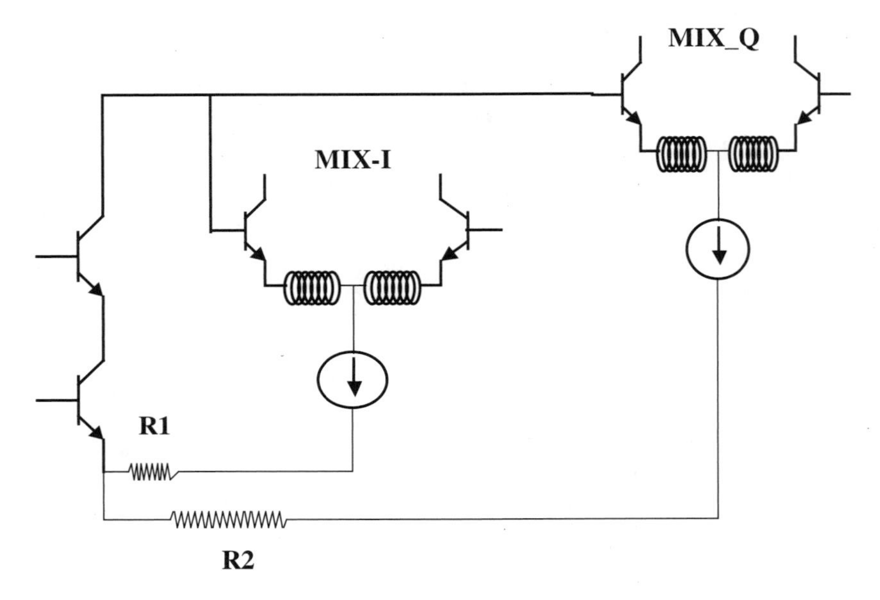


**Fig. 4.22.** Illustration of unequal feedback mechanism leading to receiver *I*/*Q* imbalance.

amenable to deviation from its optimization point when fabricated. This contributes to some extent, as there exists some amount of *I*/*Q* imbalance in the micromixer measurement results (0.25 dB amplitude imbalance, 2.4 degree phase imbalance; see Table 4.5). As the same circuit was used for the frequency divider in both the cases, it would give rise to some *I*/*Q* imbalance.

In the integrated receiver layout, the LNA ground and in-phase mixer (Mixer-I) grounds have been connected together and then connected to the overall chip ground. However, this scenario was not present in case of the quadrature mixer (Mixer-Q), and the LNA ground was not connected to the Mixer-Q ground. Hence, the Mixer-Q ground is further away from the LNA ground compared to the Mixer-I ground. This causes an unequal feedback mechanism between the LNA and the two mixers in quadrature. The feedback mechanism takes place through the finite resistance of the ground connection, as shown in Fig. 4.22. The resistors R1 and R2 come from the on-chip ground connection. They cause unequal positive feedbacks to the input, and since R1 < R2, the I-channel amplitude is higher than the Q-channel. To solve this problem, LNA and mixer grounds should be separately connected to the common circuit ground, and the interconnections of LNA and mixer ground in any other form should be avoided.

## 4.7. CONCLUSION

This chapter has provided the reader with a systematic approach to silicon-based receiver design techniques based on active-mixer architectures. Direct conversion architectures for IEEE 802.11a have been used as a vehicle to illustrate the intertwined world of wireless system design, circuit design, simulation, integration and layout techniques, and optimization. A whole design cycle, starting from standard specification to high-frequency receivers in commercially available silicon-based processes has been demonstrated.

However, with what we have learned up to now, let us look one step ahead. Let us ask ourselves the following questions: What are the assumptions we have used so far? What additional techniques are needed for high-frequency receiver designs? How should we tailor our thought processes? Concentrating on the material presented in this chapter so far, we can strongly argue that the 2× architecture was not an optimum choice with regard to power consumption. As engineers, let us concentrate on the thought process once more: Should we optimize the existing frequency divider? Should we look for a ½× architecture? Will a ½× architecture be sufficient to handle the next generation of ultrahigh-frequency receivers?

The interested reader at this point might realize that there exists a need to investigate techniques with which we can design receivers with very low frequency LO signal that will be capable of down-converting the ultrahigh frequency of the RF carrier down to baseband. We now understand the limits of technology and feasibility of receiver design from a practical perspective. This need opens up a new vista of subharmonic receivers, which will be discussed in detail in the next chapters.

## REFERENCES


1. B. Gilbert, "A precise four-quadrant multiplier with subnanosecond response," *IEEE Journal of Solid State Circuits, SC-9,* 365–373, December 1968.
2. J. D. Cressler, "SiGe HBT technology: A new contender for Si based RF and microwave applications," *IEEE Transactions on Microwave Theory and Techniques, 46,* 5, 2, 572–589, May 1998.
3. *http://www.ibm.com/news/2001/06/25.phtml.*
4. IEEE Draft Supplement to IEEE Std 802.11, Wireless LAN Medium Access Control (MAC) and Physical Layer (PHY) Specifications: High-speed Physical Layer in the 5 GHz Band, September 1999.
5. H. Samavati, H. R. Rategh, and T. H. Lee, "A 5 GHz CMOS wireless LAN receiver front end," *IEEE Journal of Solid State Circuits, 35,* 5 765–772, May 2000.
6. B. Razavi, "A 5.2 GHz CMOS receiver with 62-dB image rejection," in *2000 Symposium on VLSI Circuits, Digest of Technical Papers,* pp. 34–37.
7. B. Gilbert, "The MICROMIXER: A highly linear variant of Gilbert mixer using a bisymmetric Class-AB input stage," *IEEE Journal of Solid State Circuits, 32,* 9 1412–1423, September 1997.
8. B. Gilbert, "The Multi-tanh principle: A tutorial overview," *IEEE Journal of Solid State Circuits, 33,* 1 2–16, January 1998.
9. K. Lim, A. Obatoyinbo, A. Sutono, S. Chakraborty, C.-H. Lee, E. Gebara, A. Raghavan, and J. Laskar, "A highly integrated transceiver module for 5.8 GHz OFDM Communication System using Multi-layer Packaging Technology," in *Microwave Symposium Digest, 2001 IEEE MTT-S International,* Vol. 3, pp. 1739–1742.

# 5

# SUBHARMONIC RECEIVER DESIGNS

## INTRODUCTION

Having introduced the basic receiver front-end design process in the last chapter, we now focus on subharmonic-mixer-based receiver architectures. In the history of subharmonic receiver development, passive-mixer-based techniques preceded their active counterparts. In this chapter, we provide the reader with an overview of the passive mixing technique, along with the details of practical implementation in receiver front-ends. The first part of this chapter is focused on the development of a receiver front-end in gallium arsenide (GaAs) MESFET process technology. In the second part, we focus on the implementation of passive-mixing-based techniques in silicon-based process technologies. Then we introduce the reader to the feasibility of using such techniques in very high frequency receiver front-end designs. The passive-mixer-based architectures are mostly focused on antiparallel diode pair (APDP) topology. Historically, the APDP topology was developed using Schottky barrier diodes and has become quite attractive in recent times.

In this chapter, we would like to introduce the reader to the development of the key element (Schottky barrier diode) for such receivers. A rectifying metal–semiconductor contact is known as Schottky barrier after W. Schottky, who first proposed a model for barrier formation. The knowledge of metal–semiconductor diodes is more than a century old. F. Braun in 1874 reported the rectifying nature of metallic contacts on copper, iron, and lead sulphide crystals. Point contact diodes, which employed a sharpened metallic wire in contact with an exposed semiconductor surface, were used as radio wave detectors in the early days of wireless telegraphy. They were replaced by vacuum diodes in the early 1920s.

*Modern Receiver Front-Ends.* By J. Laskar, B. Matinpour, and S. Chakraborty
ISBN 0-471-22591-6 © 2004 John Wiley & Sons, Inc.

During the Second World War, point contact diodes became important as frequency converters and low-level microwave detectors. However, point contact rectifiers proved to be very unreliable in their characteristics, and were subsequently replaced by rectifiers created by deposition of thin metallic film on a properly prepared semiconductor surface. During the Second World War, Bethe proposed the phenomenon of thermionic emission as the means of current transport over the Schottky barrier. A more detailed illustration of Schottky barrier diodes can be found in [1].

During the 1960s, more research activities focused on the Schottky barrier diode due to the advancements in metallic contacts in semiconductor technology. Further developments continued during the 1970s toward industrial production of Schottky barrier diodes. Between 1970 and 1980, several circuits were developed for microwave mixing techniques. However, these were mostly prominent in the very high frequency regime. During the 1990s, the application of Schottky-barrier-diode-based mixers became very popular in cell phone and other wireless applications, and at present they are an attractive option for high-frequency mixing for compact receiver architectures.

## 5.1. ILLUSTRATION OF SUBHARMONIC TECHNIQUES

Subharmonic mixers are key to the success of multigigahertz transceiver subsystems. The purpose of these mixers is down-conversion or up-conversion with a much lower frequency of the local oscillator (LO) tone, which is quite beneficial, as it requires generation of lower-frequency signals on-chip. These types of mixers are generally categorized by the following equation

$$\omega_{\mathrm{RF}} = n\omega_{\mathrm{LO}} + \omega_{\mathrm{IF}}$$

where $\omega_{\mathrm{RF}}$ is the RF frequency, $\omega_{\mathrm{LO}}$ is the LO frequency, $\omega_{\mathrm{IF}}$ is the IF frequency, and $n$ is an integer. For a much lower on-chip frequency generation of the LO tone, we would like $n$ to be as high as possible. However, the value of $n$ is determined by the following factors:

1. Circuit consideration. This is entirely dependent on how the output current of the mixer core is generated by the nonlinear elements. For example, the antiparallel diode pair (APDP) mixer cores [2] work in the even order harmonics of the LO tone, and, thus, $n$ is an even integer in that case. This is the reason antiparallel diode pair (APDP) mixers are also referred to as even harmonic mixers. For other cases of subharmonic mixers, $n$ could be a different integer. However, $n$ is a measure of how low a LO frequency one can generate on-chip.
2. System consideration. This is dependent on the in-band component of the LO tone. As $n$ is an integer, some of the LO harmonics eventually falls in-band.

In a direct conversion receiver, this generates some level of DC offset in the receiver. A higher value of $n$ indicates lower in-band LO, thus giving rise to lower DC offset.

It is evident from the system perspective that we would like $n$ to be as high as possible. However, it will be shown later that increasing the value of $n$ leads to major compromises in other system parameters such as conversion gain (or loss) and linearity. Most of the reports to date consider second subharmonic operation, for which $n = 2$.

To date, there have been many reports of subharmonic mixing approaches [2–21]. These mixing techniques can be broadly categorized as passive and active. There are two approaches in developing a subharmonic mixing core:

1. Use of the antisymmetric transfer characteristics of nonlinear devices
2. Use of the frequency multiplier as a part of the mixer's switching core

In this chapter, we focus on the subharmonic approach governed by the antisymmetric transfer characteristics; the other technique is the focus of the next chapter. Passive subharmonic mixing could be single-diode-based or antiparallel-diode-pair-based (APDP). There have been many reports of antiparallel-diode-pair-based mixing, and in almost every case, Schottky barrier diodes have been used. However, many of the conventional silicon-based technologies do not provide a Schottky barrier diode as a technology module. This provides the motivation to investigate techniques with which one can still do passive subharmonic mixing without the use of Schottky barrier diodes, as well as investigate active subharmonic mixing techniques in silicon-based technologies.

## 5.2. MIXING USING ANTISYMMETRIC I–V CHARACTERISTICS

This section provides a brief mathematical analysis of mixing techniques using antisymmetric I–V characteristics. This provides the foundation for development of subharmonic mixing techniques for mixing techniques in multigigahertz RF frequencies. Figure 5.1 shows an example of antisymmetric I–V characteristics formed by connecting two nonlinear devices in an antisymmetric fashion. The figure can be used to derive a generic polynomial model to represent the transfer characteristics. The current through individual branches is shown along with the cut in voltage (or threshold voltage, as appropriate) by $V_C$. The output current is denoted by $i_{out}$ and the applied voltage is denoted by $V$.

The current–voltage relationships are governed by the following relationships:

$$i_1 = K(V - V_t)^m \tag{5.1}$$

$$i_2 = -K(-V - V_t)^m \tag{5.2}$$

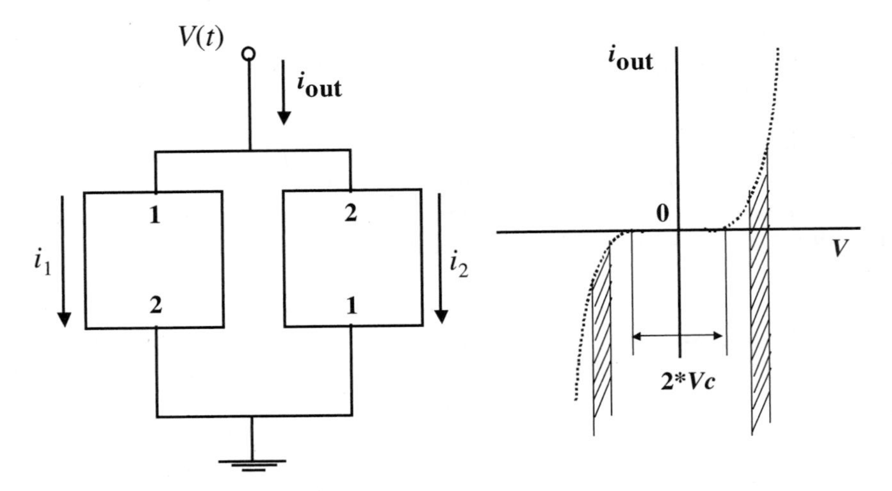


**Fig. 5.1.** Configuration of antisymmetric devices and polynomial nature of the antisymmetric characteristics.

where $K$ is a constant and $m$ is an integer. The combined transconductance is given by

$$
\begin{aligned}
g_T = g_1 + g_2 &= \frac{di_1}{dV} + \frac{di_2}{dV} \\
&= Km[\{1 + (-1)^{m-1}\}V^{m-1} + (m-1)V_t\{(-1) + (-1)^{m-1}\}V^{m-2} \\
&\quad + \frac{(m-1)(m-2)}{2}V_t^2\{1 + (-1)^{m-1}\}V^{m-3} \\
&\quad + \frac{(m-1)(m-2)(m-3)}{3!}V_t^3\{1 + (-1)^{m-1}\}V^{m-4} + \ldots]
\end{aligned}
\tag{5.3}
$$

With an input waveform of $V = V_{LO}\cos(\omega_{LO}t) + V_{RF}\cos(\omega_{RF}t)$, it could be interpreted from Eqn. 5.3 that to realize higher-order harmonic mixing, we would require the index $m$ to be an odd integer, so that we get a contribution in the output current from the very first term of the series. Also, a higher value of $m$ is desired so that we can operate our hypothetical mixer core at a higher harmonic of the LO frequency.

Now let us investigate practical examples of the polynomial nature of I–V characteristics. A natural interpretation is to apply the above formulation to square law as well as exponential characteristics. Using a very simplistic expression in case of the square law devices, we obtain

$$
i_1 = \frac{\beta_n}{2}(V_{GS} - V_t)^2(1 + \lambda V_{DS}) = \frac{\beta_n}{2}(V - V_t)^2(1 + \lambda V) \tag{5.4}
$$

$$i_2 = -\frac{\beta_n}{2}(-V_{GS} - V_t)^2(1 - \lambda V_{DS}) = -\frac{\beta_n}{2}(V + V_t)^2(1 - \lambda V) \tag{5.5}$$

From the above equations, we deduce the transconductance term to be

$$g_T = g_1 + g_2 = \beta_n\{(V_t^2 - 2V_t) + V^2(1 + 2\lambda)\} \tag{5.6}$$

The total output current is given by

$$\begin{aligned} i_{out} &= g_T V = g_T(V_{LO} \cos\omega_{LO}t + V_{RF} \cos\omega_{RF}t) \\ &= A \cos(\omega_{LO}t) + B \cos(\omega_{RF}t) + C \cos(3\omega_{LO}t) + D \cos(\omega_{RF} - 2\omega_{LO})t \\ &\quad + E \cos(2\omega_{RF} - 2\omega_{LO})t + F \cos(2\omega_{RF} + \omega_{LO})t \end{aligned} \tag{5.7}$$

The IF component is denoted by,

$$i_{IF} = \frac{3V_{LO}^2}{4} V_{RF}(1 + 2\lambda) \cos(\omega_{RF} - 2\omega_{LO})t \tag{5.8}$$

The term of interest lies in the component providing $(\omega_{RF} - 2\omega_{LO})$. This is usually referred to as the second subharmonic operation, and as seen, the quadratic characteristics are not capable of withstanding higher-order harmonic mixing (e.g., fourth and eighth harmonic mixing).

Having illustrated the square law transfer characteristics of nonlinear devices, let us consider exponential nonlinearity of I–V characteristics. This analysis was originally performed by M. Cohn et al. in 1975 [2]. Exponential characteristics can be realized easily with the help of diodes. In this case, we can replace the nonlinear devices shown in Fig. 5.1 by Schottky barrier diodes, leading to the following equations:

$$i_1 = -i_s(e^{-\alpha V} - 1) \tag{5.9}$$

$$i_2 = i_s(e^{\alpha V} - 1) \tag{5.10}$$

where $\alpha$ is the diode slope parameter ($\alpha \approx 38V^{-1}$ for GaAs Schottky diodes). Similar analysis to that above results in

$$g_T = g_1 + g_2 = \frac{di_1}{dV} + \frac{di_2}{dV} = \alpha i_s(e^{\alpha V} + e^{-\alpha V}) = 2\alpha i_s \cosh(\alpha V) \tag{5.11}$$

With the applied voltage of $V = V_{LO} \cos(\omega_{LO}t) + V_{RF} \cos(\omega_{RF}t)$, as before, and including only the LO waveform in the transconductance term under the assumption $V_{LO} \gg V_{RF}$, the transconductance expression in Eqn. 5.11 reduces to

$$g_T = g_1 + g_2 = 2\alpha i_s[I_0(\alpha V_{LO}) + 2I_2(\alpha V_{LO})\cos(2\omega_{LO}t) + 2I_4(\alpha V_{LO})\cos(4\omega_{LO}t) + \ldots] \tag{5.12}$$

where $I_n(\alpha V_{LO})$ are modified Bessel functions of the second kind. For the applied voltage

$$V = V_{LO}\cos(\omega_{LO}t) + V_{RF}\cos(\omega_{RF}t)$$

the output current expression is

$$i = g_T V = g_T[V_{LO}\cos(\omega_{LO}t) + V_{RF}\cos(\omega_{RF}t)] \tag{5.13}$$

Equation 5.13 results in the output current:

$$\begin{aligned} i = {} & A\cos(\omega_{LO}t) + B\cos(\omega_{RF}t) + C\cos(3\omega_{LO}t) + D\cos(5\omega_{LO}t) + E\cos(\omega_{RF} + \omega_{LO})t \\ & + F\cos(\omega_{RF} - \omega_{LO})t + G\cos(\omega_{RF} + 4\omega_{LO})t + H\cos(\omega_{RF} - 4\omega_{LO})t + \ldots \end{aligned} \tag{5.14}$$

As seen in Eqn. 5.14, the output current contains frequency terms ($m\omega_{LO} \pm n\omega_{RF}$), where $(m + n)$ is an odd integer, i.e., $(m + n) = 1, 3, 4, \ldots$. As the output current is formed by the interactions with even harmonics of the LO waveform, this type of mixer is often referred to as an "even harmonic Mixer" (EHM).

The above discussion has been fully focused on a mixing technique based on a single-tone LO signal. A potential disadvantage of this approach is that the mixing can only take place when the LO swing gets past the threshold region. Hence, process technologies providing very low threshold voltage devices are preferable

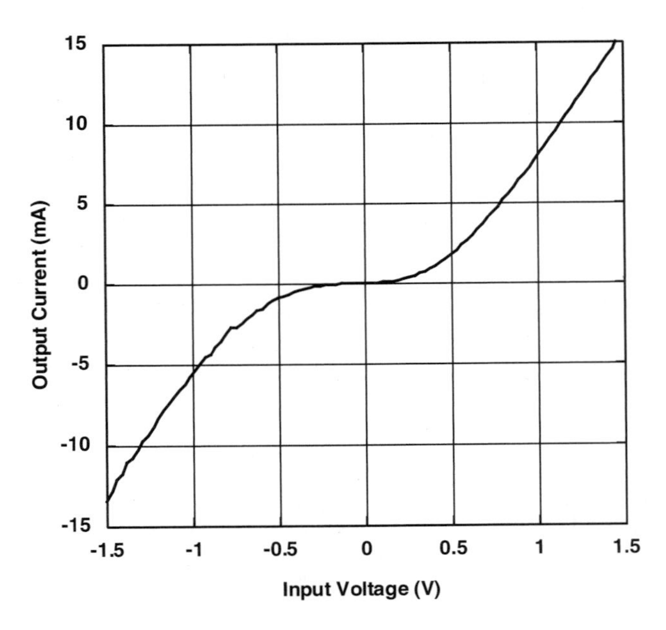


**Fig. 5.2.** Example of antisymmetric transfer characteristics in 0.5 μm SOI CMOS.

for reducing the LO voltage swing required for the mixing operation. An example is the 0.5 μm silicon-on-insulator (SOI) process, providing a threshold voltage of 0.2 V. Figure 5.2 shows the antisymmetric characteristics formed by antiparallel diodes fabricated in this process.

## 5.3. IMPACT OF MISMATCH EFFECTS

In this section, we introduce an illustration of mismatch effects and their impact on mixer performance characteristics such as DC offset, which is critical to the development of direct conversion receivers. We also introduce a technique of injecting the second harmonic tone of the LO tone to alleviate the DC offset generated by such mismatch effects in the mixer core. DC offset cancellation techniques in the RF domain are not very popular due to their low reliability, accuracy, and reproducibility, and most of the DC offset schemes are implemented in the baseband as of today. However, we demonstrate the feasibility of such techniques, using fundamental device equations. Although monolithic processes can generally reduce mismatch between adjacent diodes so that it can be ignored for most performance criteria, effects of the mismatch on the DC offset and other even-order products might prove significant. To fully understand the mechanisms of imbalance effects created by mismatch between the I–V characteristics of the diodes, we first present a theoretical analysis of such mismatch effects, and then present measurement results to validate the theory.

Using fundamental equations, we describe the mixing mechanism in an APDP and present the effects of mismatch on DC offsets, intermodulations, and LO leakage. We derive the fundamental equations of the mismatched APDP in Fig. 5.3.

Starting with Eqns. 5.9 and 5.10, we introduce mismatches between the diodes, considering the difference in both the slopes and saturation currents [2]. We define the modified slopes and saturation currents as

$$i_{s1} = i_s + \Delta i_s \qquad \text{and} \qquad i_{s2} = i_s - \Delta i_s \tag{5.15}$$

and

$$\alpha_1 = \alpha + \Delta\alpha \qquad \text{and} \qquad \alpha_2 = \alpha - \Delta\alpha \tag{5.16}$$

Then we rewrite the instantaneous conductance expressions for each case:

$$g_{\Delta i_s} = 2\alpha i_s \left\lfloor \cosh(\alpha V) + \frac{\Delta i_s}{i_s} \sinh(\alpha V) \right\rfloor \tag{5.17}$$

and

$$g_{\Delta\alpha} = 2\alpha i_s e^{(\Delta\alpha)V} \left\lfloor \cosh(\alpha V) + \frac{\Delta\alpha}{\alpha} \sinh(\alpha V) \right\rfloor \tag{5.18}$$

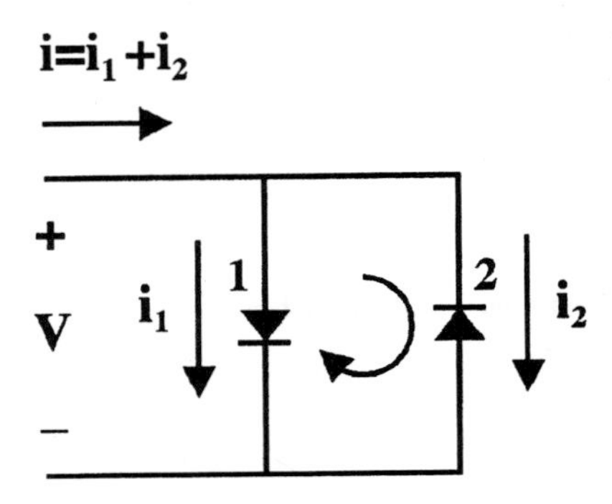


**Fig. 5.3.** Schematic of the antiparallel diode pair.

where $V = V_{LO}\cos(\omega_{LO}t)$. Now, two modified expressions can be written to describe the change in conductance due to mismatch in each variable:

$$g_{\Delta is} \approx 2\alpha i_s[I_0(\alpha V_{LO}) + 2I_2(\alpha V_{LO})\cos(2\omega_{LO}t) + 2I_4(\alpha V_{LO})\cos(4\omega_{LO}t) + \ldots] \\ + 2\alpha(\Delta i_s)[2I_1(\alpha V_{LO})\cos(\omega_{LO}t) + 2I_3(\alpha V_{LO})\cos(3\omega_{LO}t) + \ldots] \tag{5.19}$$

and

$$g_{\Delta\alpha} \approx 2\alpha i_s[I_0(\Delta\alpha V_{LO}) + I_1(\Delta\alpha V_{LO})\cos(\omega_{LO}t) + I_1(\Delta\alpha V_{LO})\cos(2\omega_{LO}t) + \ldots] \\ \times \{[I_0(\alpha V_{LO}) + 2I_2(\alpha V_{LO})\cos(2\omega_{LO}t) + \ldots] \\ + (\Delta\alpha/\alpha)[2I_1(\alpha V_{LO})\cos(\omega_{LO}t) + \ldots]\} \tag{5.20}$$

where $I_n(x)$ denotes the $n$th-order modified Bessel function. These conductance terms can then be simplified for the desired baseband output, DC offset, intermodulations, and LO leakage components.

It should be noted that accurate results are dependent on the load impedance present at the diodes at various critical frequencies such as the fundamental mixing frequency and its second harmonic. Therefore, with the exception of DC offset, these formulae derived from the above expressions can only be used for conceptual understanding of the mixing mechanism and not for predicting actual performance. In the case of the DC offset, the formulae are less susceptible to loading effects and can be used for estimating this particular product. DC offset has no relation to the load at the mixing frequency or its harmonics, but only depends on the LO voltage and the zero-frequency load impedance of the APDP.

To further understand the mixing mechanism, we solve for different mixing products by manipulating the conductance expressions of Eqns. 5.19 and 5.20. After simplification, we collect all the significant conductance terms for each product. For DC offset we can write

$$g_{\Delta i_s,\,\text{DC offset}} \approx 2\alpha(\Delta i_s)I_1(\alpha V_{\text{LO}})\cos(\omega_{\text{LO}}t), \tag{5.21}$$

and

$$g_{\Delta\alpha,\,\text{DC offset}} \approx 2\alpha i_s \cos(\omega_{\text{LO}}t)I_1(\Delta\alpha V_{\text{LO}}) \times [I_0(\alpha V_{\text{LO}}) + I_2(\alpha V_{\text{LO}})]. \tag{5.22}$$

Current characteristics of the APDP can then be found by multiplying the conductance expressions by the applied voltage

$$V = V_{\text{LO}}\cos(\omega_{\text{LO}}t) + V_{\text{RF}}\cos(\omega_{\text{RF}}t)$$

and summing the two currents,

$$i_{\text{DC offset}} \approx \pm 2\alpha(\Delta i_s)V_{\text{LO}}I_1(\alpha V_{\text{LO}}) + 2\alpha i_s V_{\text{LO}}I_1(\Delta\alpha V_{\text{LO}}) \times [I_0(\alpha V_{\text{LO}}) + I_2(\alpha V_{\text{LO}})] \tag{5.23}$$

Current terms add constructively when one of the diodes has both a higher slope and a higher saturation current; otherwise, they add destructively. A careful look at the initial diode current characteristics described in Eqns. 5.9 and 5.10 show that the existence of the former condition is unlikely. Higher saturation current and higher slope will result in considerable difference in the I–V characteristics of the diodes in the APDP, a condition that is not common in most monolithic processes. A more likely scenario is the latter when one of the diodes has a higher slope and a lower saturation current. Therefore, Eqn. 5.23 can be rewritten as

$$i_{\text{DC offset}} \approx +2\alpha(\Delta i_s)V_{\text{LO}}I_1(\alpha V_{\text{LO}}) - 2\alpha i_s V_{\text{LO}}I_1(\Delta\alpha V_{\text{LO}}) \times [I_0(\alpha V_{\text{LO}}) + I_2(\alpha V_{\text{LO}})] \tag{5.24}$$

Similarly, current characteristics of the APDP are found for loss, in-band (2LO) leakage, and second-order products:

$$i_{\text{out}} \approx \left| \begin{array}{l} +2\alpha i_s V_{\text{RF}} I_2(\alpha V_{\text{LO}}) + \ldots \\ -2\alpha i_s V_{\text{RF}} \times [I_0(\Delta\alpha V_{\text{LO}})I_2(\alpha V_{\text{LO}})] - \ldots \\ -2(\Delta\alpha) i_s V_{\text{LO}} \times [I_1(\Delta\alpha V_{\text{LO}})I_1(\alpha V_{\text{LO}})] \end{array} \right| \times \cos(2\omega_{\text{LO}}t - \omega_{\text{RF}}t) \tag{5.25}$$

and

$$i_{2\text{LO}} \approx \left| \begin{array}{l} +2\alpha(\Delta i_s)V_{\text{LO}} \times [I_1(\alpha V_{\text{LO}}) + I_3(\alpha V_{\text{LO}})] + \ldots \\ -2(\Delta\alpha) i_s V_{\text{LO}} I_0(\Delta\alpha V_{\text{LO}})I_1(\alpha V_{\text{LO}}) - \ldots \\ -2\alpha i_s V_{\text{LO}}[I_1(\Delta\alpha V_{\text{LO}})I_2(\alpha V_{\text{LO}}) + I_1(\Delta\alpha V_{\text{LO}})I_0(\alpha V_{\text{LO}})] \end{array} \right| \times \cos(2\omega_{\text{LO}}t) \tag{5.26}$$

and

$$i_{\mathrm{IM2}} \approx \left| \begin{matrix} +2\alpha(\Delta i_s)V_{\mathrm{RF}}I_1(\alpha V_{\mathrm{RF}}) + \ldots \\ -2\alpha i_s V_{\mathrm{RF}}I_1(\Delta\alpha V_{\mathrm{RF}}) \times [I_0(\alpha V_{\mathrm{RF}}) + I_2(\alpha V_{\mathrm{RF}})] \end{matrix} \right| \times \cos(\Delta\omega_{\mathrm{RF}}t) \quad (5.27)$$

where $\Delta\omega_{\mathrm{RF}}$ is the difference between two in-band RF tones with the same power.

By examining Eqns. 5.25 and 5.26, it becomes clear that baseband output, 2LO, and DC offset are all dependent on the LO amplitude, $V_{\mathrm{LO}}$, the LO voltage applied to the APDP. In addition, the IM2 product also has a dependence on $V_{\mathrm{LO}}$. The LO amplitude, $V_{\mathrm{LO}}$, also has an indirect impact on the IM2 value by affecting the values extracted for $\alpha$ and $i_s$. This is further explained in the next section.

Upon further examination of Eqn. 5.27, we also see that, interestingly enough, there is a dependence on $V_{\mathrm{RF}}$, the input RF power. This phenomenon can explain the typical differences between the theoretical 2:1 slope and the measured slope of IM2 versus RF input power in a two-tone test.

## 5.4. DC OFFSET CANCELLATION MECHANISMS

In this section, we use analytical formulations developed in the previous section to explore several cancellation mechanisms for DC offset and even-order intermodulation distortions.

### 5.4.1. Intrinsic DC Offset Cancellation

As described in the last section, the DC offset current of the mismatched APDP can be described by Eqn. 5.24 as

$$i_{\mathrm{DC\ offset}} \approx |+2\alpha(\Delta i_s)V_{\mathrm{LO}}I_1(\alpha V_{\mathrm{LO}}) - 2\alpha i_s V_{\mathrm{LO}}I_1(\Delta\alpha V_{\mathrm{LO}}) \times [I_0(\alpha V_{\mathrm{LO}}) + I_2(\alpha V_{\mathrm{LO}})]|$$

A second look at the above equation uncovers an interesting characteristic. Depending on the parameters such as $V_{\mathrm{LO}}$, $i_s$, $\Delta i_s$, $\alpha$, and $\Delta\alpha$, both negative and positive DC offsets can be generated. More importantly, under special conditions, it is possible for the two terms in the equation to cancel each other and produce a net DC offset of zero. Consequently, this mechanism not only eliminates DC offset generated by mismatch but also cancels any DC offset generated by other sources such as LO self-mixing. As shown in Eqn. 5.24, the intrinsic DC offset cancellation mechanism is directly related to the LO voltage applied across the APDP. Therefore, adjusting the LO power can vary the level of DC offset generated by the mismatch to produce a net DC offset of zero, a condition referred to as a DC offset null. While operating at this null point, all concerns regarding saturation of the direct-coupled baseband amplifiers is eliminated and any unexpected formation of DC offset can be compensated for by minor tuning of the LO power.

A similar intrinsic cancellation mechanism can also be induced for the 2LO and

IM2 extraneous signals described in Eqns. 5.26 and 5.27:

$$i_{2\mathrm{LO}} \approx \left| \begin{array}{l} +2\alpha(\Delta i_s)V_{\mathrm{LO}} \times [I_1(\alpha V_{\mathrm{LO}}) + I_3(\alpha V_{\mathrm{LO}})] + \dots \\ -2(\Delta\alpha)i_s V_{\mathrm{LO}} I_0(\Delta\alpha V_{\mathrm{LO}}) I_1(\alpha V_{\mathrm{LO}}) - \dots \\ -2\alpha i_s V_{\mathrm{LO}}[I_1(\Delta\alpha V_{\mathrm{LO}}) I_2(\alpha V_{\mathrm{LO}}) + I_1(\Delta\alpha V_{\mathrm{LO}}) I_0(\alpha V_{\mathrm{LO}})] \end{array} \right| \times \cos(2\omega_{\mathrm{LO}} t)$$

and

$$i_{\mathrm{IM2}} \approx \left| \begin{array}{l} +2\alpha(\Delta i_s)V_{\mathrm{RF}} I_1(\alpha V_{\mathrm{RF}}) + \dots \\ -2\alpha i_s V_{\mathrm{RF}} I_1(\Delta\alpha V_{\mathrm{RF}}) \times [I_0(\alpha V_{\mathrm{RF}}) + I_2(\alpha V_{\mathrm{RF}})] \end{array} \right| \times \cos(\Delta\omega_{\mathrm{RF}} t)$$

Similarly, the LO power can be manipulated to obtain a null point for IM2 and 2LO. However, it is unlikely to reach a condition in which all three components (DC offset, IM2, and 2LO) simultaneously experience a null point. This minimizes the use of this mechanism for only one of the extraneous components of the mixer output and leads to the need to prioritize the design objectives. Our experimental results described in the next section show that the in-band 2LO generation in the APDP remains minimal for monolithic structures. This eliminates the in-band interference of 2LO, leaving only DC offsets and IM2 to affect the receiver performance.

Another drawback of this intrinsic cancellation mechanism results from the need for LO tuning and the direct relationship between LO power and conversion loss. A serious problem can arise when the nulls only occur for the LO powers that are either too low and result in poor conversion characteristics, or too high and exceed the diode current limitations.

This encourages us to explore other methods of cancellation with greater flexibility that can compliment this intrinsic technique and allow for simultaneous cancellation of both IM2 and DC offset. In the next section, we introduce a proposed extrinsic method of cancellation of DC offsets that provides such ability.

### 5.4.2. Extrinsic DC Offset Cancellation

In order to eliminate the DC offsets with greater control, reproducibility, and flexibility, we propose an extrinsic cancellation method that uses a canceling DC component generated by down-converting an additional tone at the second harmonic of the LO, $V_{2\mathrm{LO}} \cos(\omega_{2\mathrm{LO}} t + \phi)$. The DC current generated from the addition of this tone can be described by multiplying the conductance expressions in Eqns. 5.19 and 5.20 with the applied voltage and collecting the significant terms that arise from mixing the LO signal and the second harmonic,

$$i_{\mathrm{DC\ cancel}} \approx V_{\mathrm{LO}} V_{2\mathrm{LO}} \cos(\phi)\alpha i_s [I_2(\alpha V_{\mathrm{LO}}) + I_2(\Delta\alpha V_{\mathrm{LO}}) I_2(\alpha V_{\mathrm{LO}}) + \dots] \qquad (5.28)$$

Equation 5.28 shows that modifying the phase and amplitude of the second harmonic varies the amplitude and sign of this cancelling DC component; therefore, it can be controlled to produce a net DC offset of zero by matching the amplitude and countering the sign of the preexisting DC offset. Since the RF properties of on-chip active components can be controlled with relative ease and accuracy, this indirect RF tuning of the DC output can be performed with greater precision than low-frequency analog tuning.

To fully explore the practicality of this cancellation technique, we also consider the noise contribution of the second harmonic tone to the mixer noise floor and flicker noise. As shown in Fig. 5.4, the phase noise associated with the second harmonic tone is down-converted to baseband in the mixing process.

This addition of phase noise can be modeled by additive flicker noise at the output of the mixer, where the corner frequency and the shape of this flicker noise are determined by the skirt of the second harmonic tone. The signal-to-noise ratio (S/N) of signals below this corner frequency is degraded by the addition of the down-converted phase noise. Assuming a practical on-chip oscillator phase noise of −100 dBc/Hz at 100 kHz offset from the carrier, corner frequencies less than 100 kHz can be achieved. This number can decrease dramatically when a phase lock loop (PLL) synthesized source is used, which is common for most wireless applications [7].

Using the extrinsic method allows us to perform DC offset cancellation by tuning a 2LO signal while tuning the LO signal for IM2 cancellation. Therefore, a simultaneous cancellation of both signals is possible. However, it is clear that a well-designed control mechanism is required to synchronize such operations.

## 5.5. EXPERIMENTAL VERIFICATION OF DC OFFSET

In order to verify the fundamental equations derived for DC offset in the APDP, we characterized a monolithic APDP core fabricated in a commercially available GaAs MESFET process. The diodes are first measured and modeled so that the proper

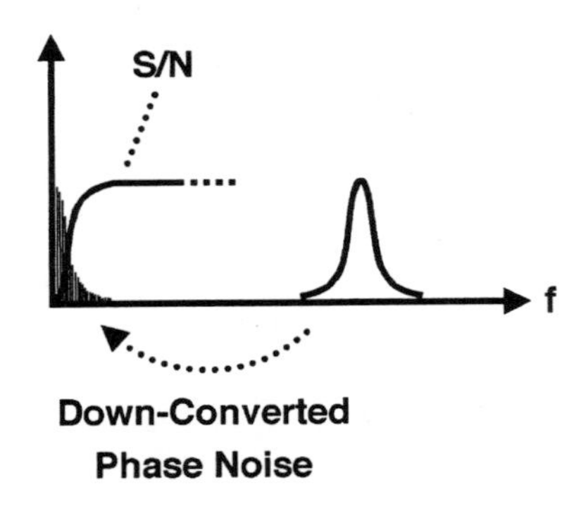


**Fig. 5.4.** Additive flicker noise effect in extrinsic DC offset cancellation technique.

variables can be extracted for use in the formulas. As shown in Fig. 5.5, multiple values for the slope and saturation current of each diode are extracted for small ranges of the junction voltage and used to accurately model the current in each bias range.

For voltages above the turn-on voltage, the diode behavior is found to be dominated by the series resistance and can no longer be described accurately by the ideal diode equation. Therefore, in this case, the validity of the formulas is limited to the "low" LO power levels of approximately below 4 dBm. The diode parameters corresponding to "low" LO voltages are extracted and substituted in Eqn. 5.24 to calculate the DC offset generated by the mismatch. The peak measured LO voltage is used to identify the appropriate range of junction voltage and choose the corresponding $i_s$, $\Delta i_s$, $\alpha$, and $\Delta\alpha$ values.

The diodes are then measured using on-wafer probes in the measurement setup shown in Fig. 5.6. Extensive filtering is used in the setup to ensure sufficient rejection of the harmonics of the LO synthesizer, which can result in extraneous DC components. The LO signal is applied to the APDP test structure while the baseband output is monitored through a bias tee.

Figure 5.7 shows the agreement between the calculated and measured values for DC offset at LO power levels varying from –4 to +4 dBm. In-band 2LO leakage of less than –50 dBm is also measured over the LO power range, a low enough level that will not impact the DC offset results by a second down-conversion in the APDP. The measurements verify the validity of the fundamental formulas in predicting the DC offset generated under "low" LO conditions. Small inconsistencies between the calculated and measured values are attributed to the effect of the load impedance, which is not considered in the formulas.

The above measurement has been repeated for various APDP test structures of different peripheries to examine the feasibility of the intrinsic DC offset cancellation described earlier. The LO power is varied over a wider range of powers, reaching as high as 16 dBm. As shown in Fig. 5.8, measurements of fabricated APDP test

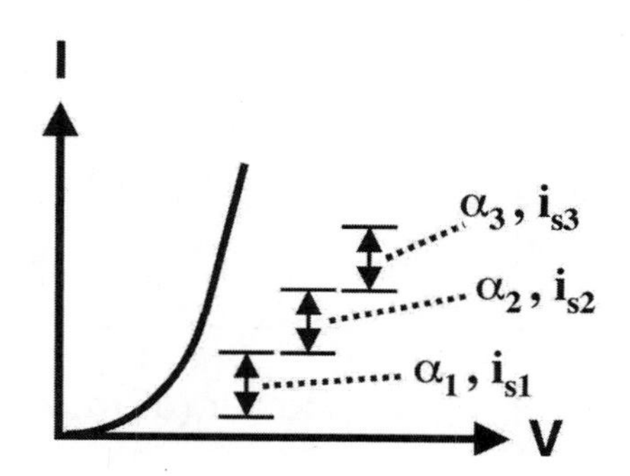


**Fig. 5.5.** Segmentation of diode I–V curve for accurate modeling.

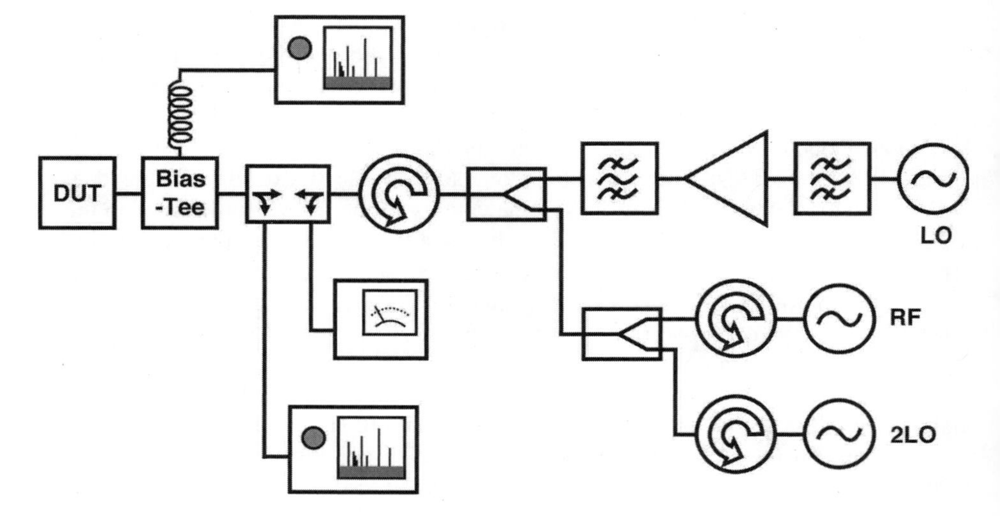


**Fig. 5.6.** Measurement setup used for the verification of APDP DC offset generation and cancellation mechanisms.

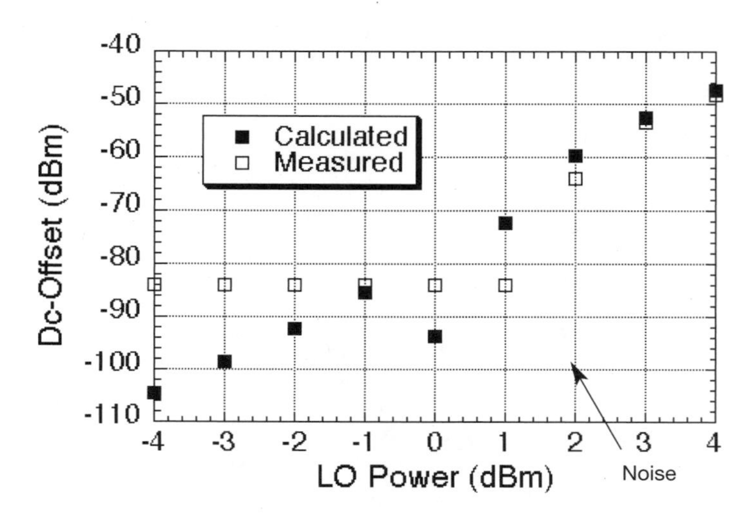


**Fig. 5.7.** Measured and calculated DC offset of a monolithic GaAs APDP at an LO frequency of 2 GHz.

structures demonstrate DC offset cancellation by the intrinsic mechanism. Measurements confirm cancellation within the LO power range that produces optimum conversion characteristics in terms of loss and intermodulations. Although these results verify the intrinsic mechanism of DC offset cancellation, further study is needed to investigate the repeatability of this technique.

An experiment was also performed to verify the proposed extrinsic DC offset cancellation technique. As shown in Fig. 5.6, two synthesized sources are used to generate the LO and second harmonic of the LO frequency. The signals are combined and applied across the APDP while the baseband output is monitored through a bias tee. With the phase kept at a constant offset, the amplitude of the second harmonic is varied by the source until DC offset cancellation is observed. Figure 5.9 shows the DC output of the APDP as a function of second harmonic power.

In a three-port test setup similar to that of Fig. 5.6, the experiment is repeated with a C-band EH direct conversion mixer that incorporates the same APDP structure. The balanced mixer consists of active baluns at the RF input and two sets of APDP mixers that utilize resonant tanks to reduce DC offsets generated by LO leakage [8].

Measurement results verify effective DC offset cancellation while maintaining a second-order input intercept point (IIP2) of +16 dBm and an average conversion loss of 10 dB for input RF frequencies of 5 to 6 GHz. Fig. 5.10 shows the conversion loss and DC offset performance of this mixer. Die photographs of the test structures are shown in Fig. 5.11.

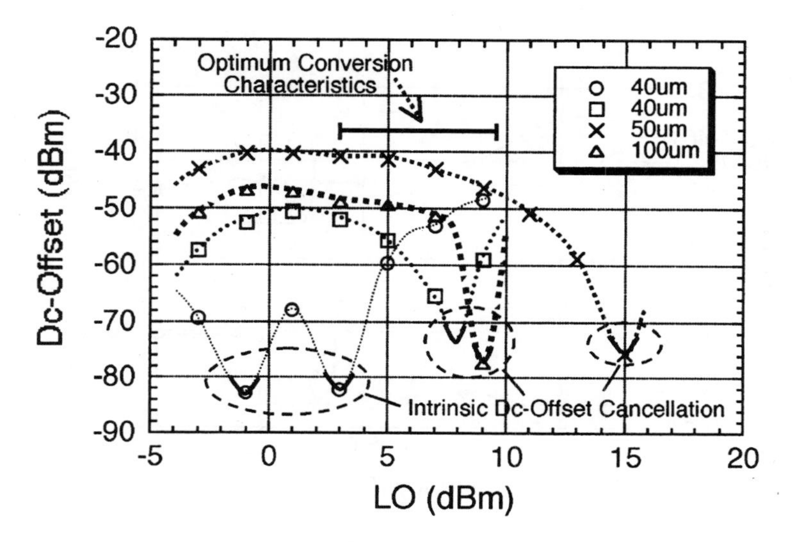


**Fig. 5.8.** DC offset nulls created by the intrinsic cancellation for different diode geometries.

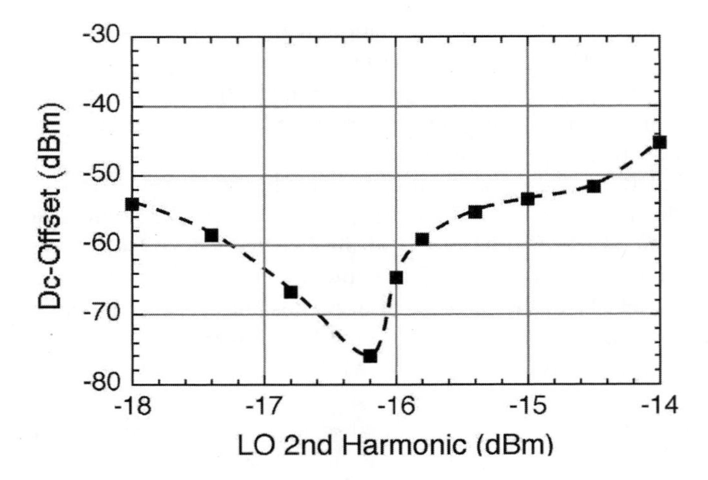


**Fig. 5.9.** DC offset cancellation by the extrinsic method as a function of the power of the second harmonic.

## 5.6. WAVEFORM SHAPING BEFORE MIXING

In the earlier sections of this chapter, we described the mixing operation using a single fundamental-tone LO signal. This mixing phenomenon is only a function of the nonlinear nature of the antisymmetric characteristics. Hence, the square law characteristics would not be able to support higher orders of subharmonic mixing, and the most feasible option is to use second subharmonic mixing in the case of antisym-

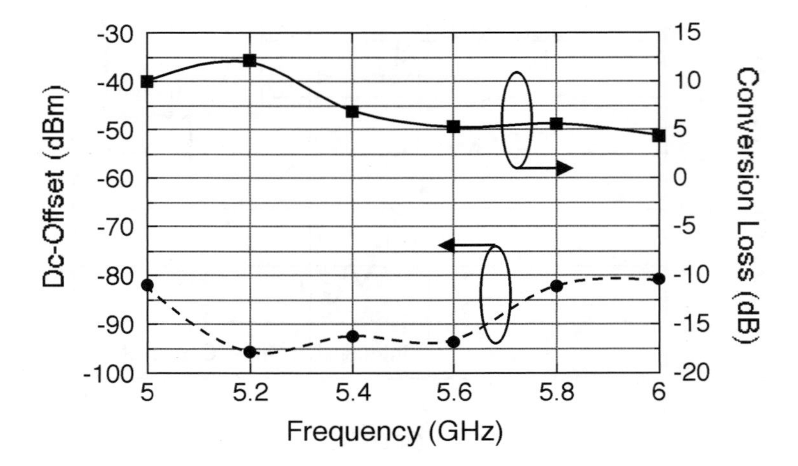


**Fig. 5.10.** DC offset and conversion loss of the EH mixer.

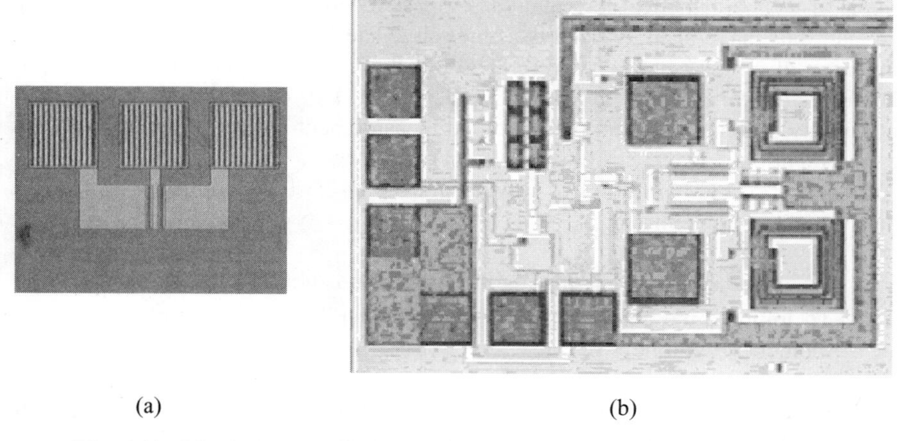

(a) (b)

**Fig. 5.11.** Die photograph of an APDP test structure and the EH mixer MMIC.

metric characteristics with square law nature. However, it is possible to shape the waveform in such a way that prior to the mixing operation we generate a number of harmonics in the LO waveform. For example, the LO waveform can be passed through a buffer to generate a square-wave-like waveform, and essentially result in the odd harmonics of the input LO signal tone. These harmonics, by the very nature of their nonlinear characteristics, would result in a sum and difference frequencies among them, thus extending the capability of the traditional square law characteristics for higher-harmonic operations. Another example is the half-wave rectifier waveform, the Fourier series expansion of which contains both the fundamental as well as a sum of even harmonics of the input frequency:

$$\begin{aligned} V_{\mathrm{LO,HWR,POS}}(t) &= \frac{V_{\mathrm{LO}}}{\pi} + \frac{V_{\mathrm{LO}}}{2}\sin(\omega_{\mathrm{LO}}t) \\ &\quad - \frac{2V_{\mathrm{LO}}}{\pi}\left\{\frac{1}{1.3}\cos(\omega_{\mathrm{LO}}t) + \frac{1}{3.5}\cos(4\omega_{\mathrm{LO}}t) + \ldots\right\} \end{aligned} \tag{5.28}$$

It is quite clear that application of the waveform shown in Eqn. 5.28 through square law characteristics or exponential characteristics would result in extension of the second subharmonic approach. In this regard, square law characteristics should also be able to withstand nonlinearity. However, when designing the signal path, one must make sure that minimum additional hardware is used.

One can observe quite interesting phenomena with regard to the injection of additional harmonic tones along with the fundamental LO. In the previous section, we described the effect of a second harmonic injection of LO tone and cancellation of even-order terms generated from mismatches in the diode pair. In this section, we will discuss the effects of the third harmonic LO tone injection in the APDP core.

Some of the mathematical formulations of these techniques can be quite complicated and, thus, have been avoided for simplicity. The complications arise in the formulation of a mathematical representation of the two-tone operation on the exponential characteristics with area-mismatched APDP core. There is also the possibility of deviation from theoretical to measurement results at higher LO powers, as mentioned in the earlier sections.

Injection of LO harmonics or waveform shaping has two implications with regard to subharmonic mixing:

1. To extend subharmonic mixing beyond second subharmonic
2. To improve the performance of the mixer while operating with higher harmonics of LO tone

We have discussed a possible scheme of waveform shaping. In the subsequent sections, we demonstrate the effects of third harmonic LO tone injection along with the fundamental. As will be shown, this scheme allows the reduction of the main subharmonic LO frequency and reduces the LO power requirement. It alleviates major problems associated with the use of fourth- and higher-order subharmonic mixers. By utilizing the odd-order harmonics of the main subharmonic LO, significant performance improvements occur while reducing the LO power.

Although the third harmonic injection technique is applicable to most subharmonic mixers, in this chapter we limit the scope of the discussion to APDP structure only. Earlier reports on APDP-based passive mixers [8, 9] utilized second subharmonic operation with regard to LO tone, which is quite suitable for low RF input frequencies (up to 6 GHz). As the RF frequency increases, on-chip integration of high-quality oscillators becomes difficult. This provides the motivation to investigate techniques that allow the mixer core to operate at higher subharmonics (lower LO frequencies). However, the conversion loss and linearity performances achieved by the application of a single-tone LO operating at fourth and eighth subharmonics are not very attractive, as their contribution is less significant compared to the second-order subharmonic LO tone [10, 11]. These performances can be greatly improved by inclusion of the third harmonic tone of the higher-order subharmonic LO frequency under application (fourth or eighth subharmonics in this case). Combining tones also helps reduce the LO power level requirement on the individual tones, thus favoring on-chip integration of such mixers for wireless transceiver applications. This technique is referred to as "third harmonic LO injection" and exhibits lower conversion loss and higher linearity compared to the higher (fourth or eighth) subharmonic operation of APDP-based passive mixers. However, the conversion loss performance is a little worse compared to the second subharmonic operation.

First, we introduce the third-order LO injection technique with a qualitative analysis. Then, the performance improvements achieved by third harmonic mixing in terms of conversion loss and linearity are presented. These performance improvements are based on an APDP core fabricated both in a commercially available GaAs MESFET technology, as well as a standard digital CMOS process. Last, the applicability of this technique to future wireless communication technologies is addressed.

### 5.6.1. Theory and Analysis

In this section, we use the formulation described earlier for a more generic harmonic mixing operation, as well as qualitative illustration of the third harmonic mixing operation. The generalized expression for the output current contributed by an APDP circuit with an input voltage waveform of $V_{LO}\cos(\omega_{LO}t) + V_{RF}\cos(\omega_{RF}t)$ can be represented by [12]

$$i_{out} = 2\alpha i_s \sum_{m=0}^{\infty}\sum_{n=0}^{\infty} F(m, n, V_{LO}, V_{RF})\cos(m\omega_{LO} \pm n\omega_{RF})t \tag{5.29}$$

A numerical example can be taken by setting values of $\omega_{LO}$ and $\omega_{RF}$ as 0.9 GHz and 1.8 GHz, respectively, for a direct conversion mixing operation. This mixing operation is termed "second-subharmonic mixing" as the frequency of the LO tone is half the input RF carrier frequency. The above concept of subharmonic mixing can be further extended in the presence of higher-order subharmonic LO tones (hence, lower LO frequency). The LO waveforms in these cases are represented by

$$V_{LO} = V_{LO1}\cos\left(\frac{\omega_{LO}}{2}t\right) \qquad \text{and} \qquad V_{LO} = V_{LO1}\cos\left(\frac{\omega_{LO}}{4}t\right)$$

respectively, in case of fourth- and eighth-subharmonic operation, assuming an LO frequency of $\omega_{LO}$ for second-subharmonic operation. Using a similar approach to that in Eqn. 5.29, the generalized expression for output current of APDP under single-tone operation is denoted by

$$i_{out} = 2\alpha i_s I_k(\alpha V_{LO})\cos\{(2\omega_{LO} - \omega_{RF})t\}$$

where $k$ denotes the subharmonic order ($k = 2, 4, 8$ for second-, fourth-, and eighth-order operation). The first few terms in the modified Bessel function series, $I_n(\alpha V_{LO})$, form a slowly decreasing sequence, with typical values of $V_{LO} = 0.6$ and $\alpha = 38$. Hence, as $k$ increases from 2 to 4 and 8 (implying fourth- and eighth-order subharmonic operation), the conversion loss performance degrades.

However, if the LO voltage waveform approaches the shape of a square wave (thus implying a higher slope of the waveform), the Schottky diodes switch more efficiently and thus both conversion loss and linearity performances improve. For a square-shaped waveform with 50% duty cycle and a period of $T$, the Fourier series expansion is given by

$$U(t) = \frac{4}{\pi}\left[\sin(\omega t) + \frac{1}{3}\sin(3\omega t) + \frac{1}{5}\cos(5\omega t) + \dots\right] \tag{5.30}$$

where $\omega = 1/T$. The first two terms are the major contributors to the square wave shape. For a more generalized study of the effects of the various harmonics involved, the fundamental and the third-order harmonic tones are combined using a power combiner. Relative amplitude variation of these different tones is also taken into consideration to study the effects of individual harmonics. These tones are easy to generate using on-chip frequency synthesizers and VCOs. Mathematically, the LO input voltage can be represented by

$$V_{\mathrm{LO}} = V_{\mathrm{LO1}} \cos(\omega_{\mathrm{LO}} t) + V_{\mathrm{LO2}} \cos(3\omega_{\mathrm{LO}} t)$$

This approach can be applied for the current subharmonic-order LO tones under consideration ($k = 2, 4, 8$), and $\omega_{\mathrm{LO}}$ would be $\omega_{\mathrm{LO}}$, $\omega_{\mathrm{LO}}/2$, and $\omega_{\mathrm{LO}}/4$, respectively. The transconductance of APDP is denoted by $g = 2\alpha i_s \cosh(\alpha V_{\mathrm{LO}})$. Solving for the above combination of waveform we obtain

$$\begin{aligned} i_{\mathrm{out}} = \alpha i_s [I_0(\alpha V_{\mathrm{LO2}}) I_k(\alpha V_{\mathrm{LO2}}) + \sum_{n=1}^{\infty} I_n(\alpha V_{\mathrm{LO2}}) \{ I_{3n-k}(\alpha V_{\mathrm{LO1}}) \\ + I_{3n+k}(\alpha V_{\mathrm{LO1}}) \} ] \cos\{(2\omega_{\mathrm{LO}} - \omega_{\mathrm{RF}})t\} \end{aligned} \tag{5.31}$$

where $k = 2, 4, 8$ respectively. Fig. 5.12 shows the improvement in transconductance obtained by third harmonic mixing compared to single-tone LO operation.

The experimental setup for third harmonic injection is similar to the one shown in Fig. 5.6. LO and RF ports have been isolated from each other. In this approach, an LO frequency is injected along with its third harmonic tone. In particular, the fundamental frequency denoted by $\omega_{\mathrm{LO}}$ is usually a subharmonic of the LO tone used for conventional even-order APDP. For even harmonic mixers for direct conversion or very low IF (VLIF) applications, $\omega_{\mathrm{LO}} = \frac{1}{2}\,\omega_{\mathrm{RF}}$. The power level of the third harmonic tone is considered to be less than or equal to that of the fundamental LO tone under consideration, to ensure the use of lower power level of the third harmonic. The difference in power levels between the fundamental and its third harmonic can be denoted by "back-off" hereafter.

Other requirements for practical implementation of the proposed technique include a broad LO power tuning range over which performance improvements should occur, and addition of minimum overhead in the system design. The technique should also allow the circuit to retain its conversion loss and linearity performances over a power tuning range of at least 2–3 dBs, otherwise the local oscillator needs to maintain extremely high levels of stability in output power. In addition, increase in back-off while retaining performance is also very attractive as this means lower power injection of the higher-frequency harmonic.

### 5.6.2. Experimental Verification on GaAs MESFET APDP

For conversion loss and linearity performances, an IF of 40 KHz has been assumed, with incoming RF frequency of 1.80004 GHz. Three different LO subharmonic

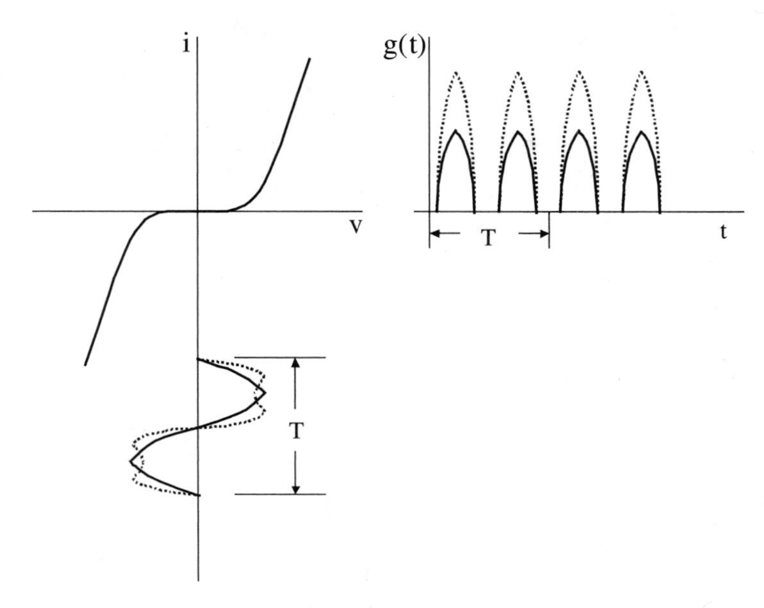


**Fig. 5.12.** Transconductance variation of APDP core with single-tone LO (solid line) and harmonic mixing (dashed line).

tones have been considered: 0.9 GHz (second subharmonic), 0.45 GHz (fourth subharmonic), and a combination of 0.45 GHz and 1.35 GHz (fourth subharmonic with third-order harmonic). LO tones at 0.45 GHz and 1.35 GHz have been considered with various back-off values and phase differences.

Figure 5.13 shows the measured conversion loss profiles as a function of LO power with a 4 dB back-off. It indicates a flattening in the conversion loss profile with the inclusion of their third harmonic tones, which is favorable for incorporation of moderate LO power variations. It is also observed from Fig. 5.13 that inclusion of third harmonic tones eliminates the nulls in the conversion loss profile, which is attractive for a broad LO tuning range operation. Also, the conversion loss performance is independent of the relative phase difference between the two tones, which has been verified with application of 45°, 90°, and 135° phase shifts between two LO tones. From Fig. 5.13, it is seen that the proposed technique improves the conversion loss by 4 dB, while reducing the LO power level by 5 dB with a back-off of 4 dB for the third harmonic tone. These performances become better with a lower back-off value, indicating an increase in the power level of third harmonic LO tone. However, the optimum conversion loss is 2 dB lower than that achieved with second-harmonic operation.

Figure 5.14 shows the measured conversion gain profiles as a function of input RF power for various configurations at optimum LO power levels for conversion gain. It is observed that the input P1dB improves by approximately 3 dB compared

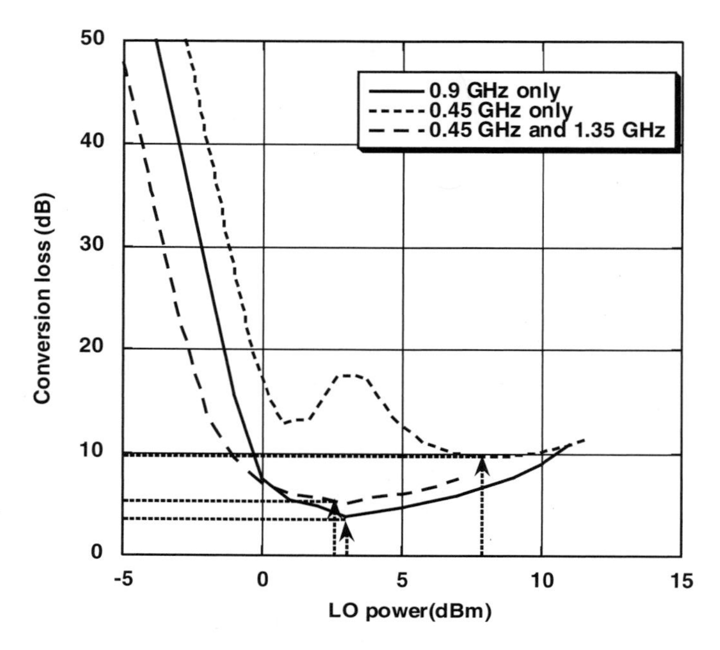


**Fig. 5.13.** Measured conversion loss versus LO power level (GaAs MESFET APDP core).

to the fourth subharmonic operation. Input P1dB increases further with reduction of back-off, similar to conversion loss performance.

As the diodes switch rapidly, the switching distortions decrease and linearity improves with the third-harmonic injection technique. Figure 5.14 illustrates the performance improvements in conversion loss and output P1dB as a function of different back-off levels in the case of fourth subharmonic (0.45 GHz) operation. These performance improvements are relative to those achieved with only a single tone LO operating at the fourth subharmonic. As the back-off increases, the effect of additional tone the third harmonic decreases and the performances tend to be similar to those with single-tone operation.

To summarize, the advantages of the third-harmonic mixing technique include:

1. Lower-frequency operation of the LO tone (fourth subharmonic or higher)
2. Improvement in loss, and linearity at a lower LO power level compared to second-subharmonic operation
3. Easy generation of additional tone to perform mixing, as it is a harmonic tone of the LO frequency
4. No performance dependence on phase difference between the two tones under consideration

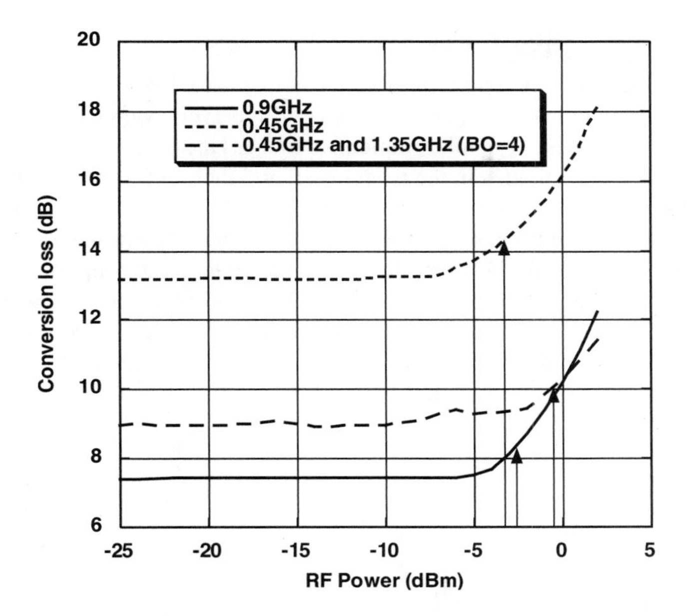


**Fig. 5.14.** Measured conversion loss versus input RF power for GaAs MESFET APDP core.

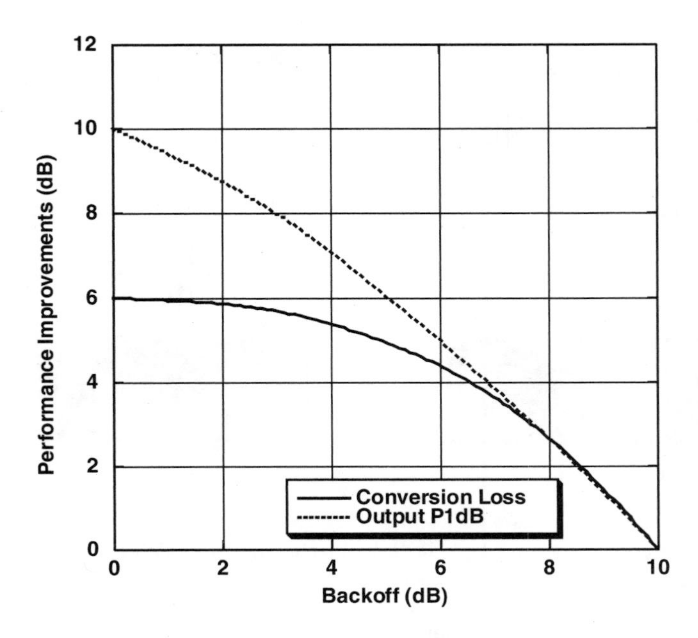


**Fig. 5.15.** Measured performance improvements achieved with harmonic injection versus various back-off values (GaAs MESFET APDP core).

### 5.6.3. Implementation in Silicon

Having illustrated the operation of Schottky-barrier-based APDP cores for generic subharmonic mixing and design of APDP-based receivers, we now concentrate on silicon based processes for similar implementations. Due to the tremendous growth of silicon-based technologies in the recent past, and their potential in fully integrated solutions, it is quite attractive to investigate the feasibility of the described techniques in silicon-based processes. Most of the commercially available silicon-based processes do not provide Schottky barrier diodes. Hence, the options are to use NWELL-based diodes, transistors configured in the form of diodes. All of these diodes utilize some sort of junction and transport of minority carrier, and junction capacitance is a major drawback in very high frequency operation.

Figure 5.16 shows the different diodes under consideration in a standard CMOS

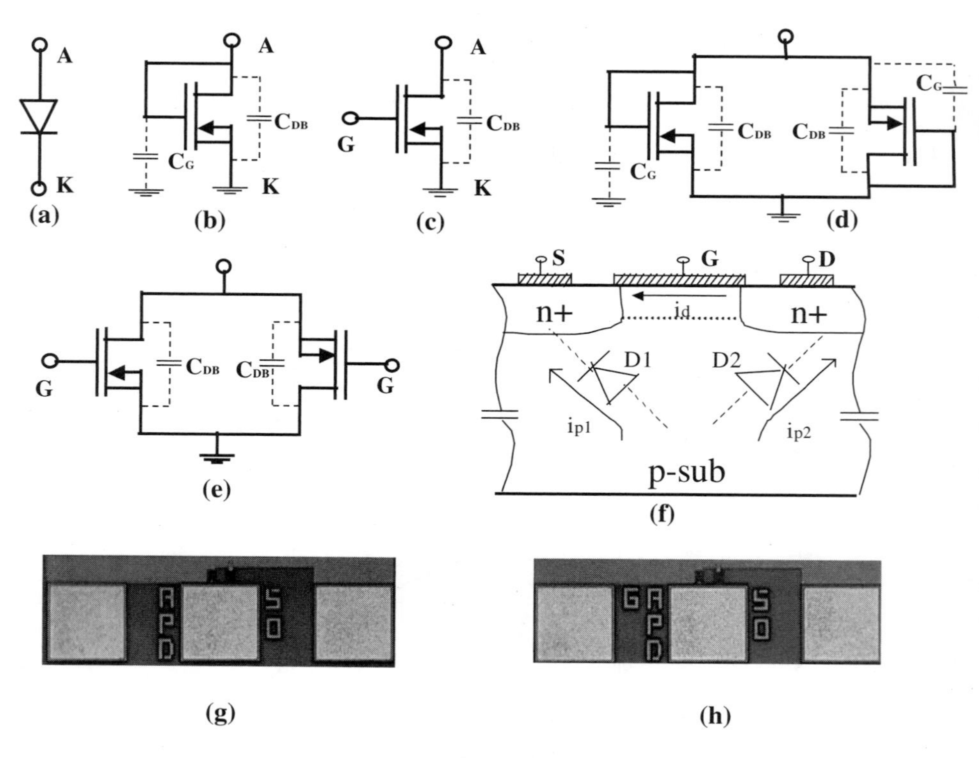


**Fig. 5.16.** Different diodes available in CMOS technology. (a) NWELL diodes. (b) NMOS diode with gate and drain shorted. (c) NMOS diode with gate open. (d) APDP core based on gate–drain-shorted MOS transistors. (e) APDP core based on open-gate MOS transistors. (f) Current flow in NMOS transistor with parasitic effects. (g) Fabricated NMOS APDP core with G–D shorted. (h) Fabricated NMOS APDP core with open gate in a 0.24 μm CMOS process.

process. Figure 5.16(c) shows a diode configuration with its gate terminal open. This diode pair topology has the advantage of minimizing parasitic capacitance, as the only parasitic capacitance is contributed by $C_{DB}$ and $C_{SB}$, as appropriate. In this section, we present the performance of the different diodes available in a standard digital CMOS process of 0.24 μm channel length. The key to this development is the source and bulk connection on the NMOS transistors. With the bulk connected to ground, the APDP core shown in Fig 5.16(d) would be the only valid one, and the resulting antisymmetric I–V characteristics would exhibit a square law nature. With the source and bulk terminals connected, the parasitic diodes play a very important role, and the resulting characteristic exhibits exponential characteristics, as explained in the following section.

Figure 5.16(f) shows the parasitic junction diodes and two different modes of current flow in an NMOS diode. When a negative voltage is applied to the drain terminal, D2 is forward-biased, D1 is at cut off as the source and bulk terminals are both at ground potential, and the current contributions from $i_{p2}$ dominate. With a positive voltage applied to the drain terminal, D2 is reverse-biased. In the case of a gate–drain-shorted NMOS transistor, applying a positive voltage to the drain terminal would cause the current ($i_d$) to follow square law characteristics. However, for an open-gate NMOS transistor, no square law dependence takes place, and only the leakage current is present in the forward-biased region. This is illustrated in Fig. 5.17.

Figure 5.18 shows the I–V characteristics of different APDP configurations

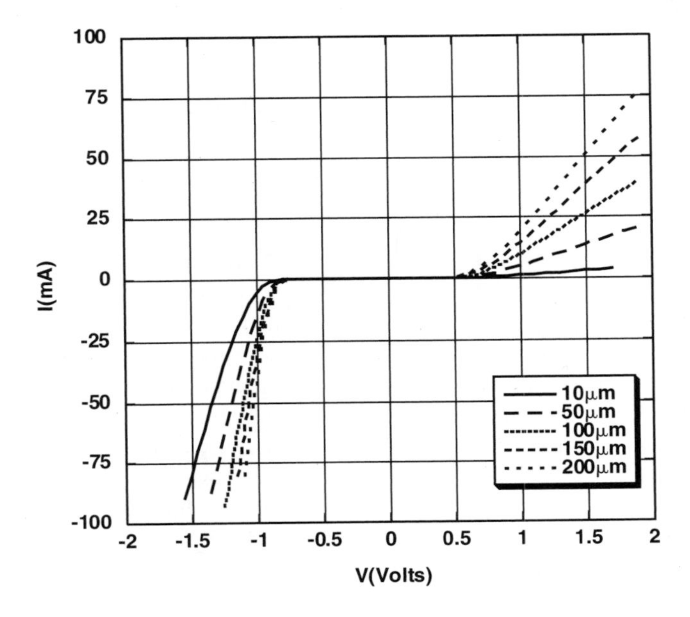


**Fig. 5.17.** Measured DC I–V characteristics of NMOS diodes of varying width with gate and drain terminals shorted.

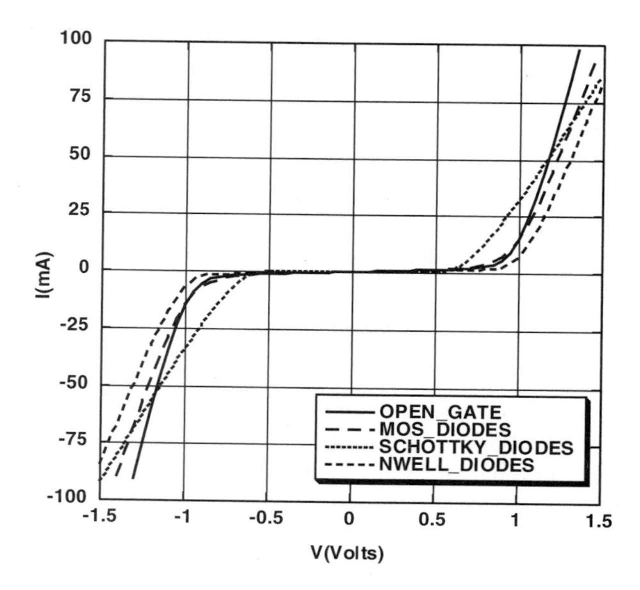


**Fig. 5.18.** Measured DC I–V characteristics of different APDP cores available in CMOS and GaAs MESFET technologies.

based on CMOS and GaAs MESFET process technologies. It is observed that, in the case of gate- and drain-connected NMOS transistors, the combined APDP characteristics are governed by an exponential relationship and the effects of the square law relationships on these diodes are dominated by the p–n+ junctions formed by p-type bulk and n+ doping regions. An exponential relationship is preferred for implementation of the third harmonic mixing technique as the conversion loss performance is governed by a slowly decreasing Bessel series. Hence, the fourth-subharmonic terms are comparable (although reduced) to second-subharmonic terms at the output.

Figure 5.18 also shows that there is little difference in the I–V characteristics for the open-gate and the gate–drain-shorted NMOS transistor APDP cores. The current in the core is based on the p–n junction formed in the bulk and n+ diffusion regions. In the case of gate–drain-shorted transistor, one NMOS transistor exhibits square law current when the other exhibits exponential current and vice-versa. These two currents flow in opposite directions, thereby reducing the overall output current. In case of an open-gate NMOS APDP core, when one transistor exhibits diode-like characteristics, the other exhibits only the leakage current. Thus, the current level is higher in open-gate topology. Also, at very high frequencies, the MOS capacitances play a major role in determining the RF performance of the APDP core. It is also observed from Fig. 5.16 that the parasitic capacitances are given by $C_p = 2(C_G + C_{DB})$ for gate–drain-shorted NMOS APDP, and $C_p = 2C_{DB}$ for the case of the open-gate NMOS APDP. A major part of the parasitic capacitance is contributed by the gate (as $C_G > C_{DB}$), and this causes a MOSFET APDP to exhibit a low-input impedance at high frequencies. With the gate open, the gate capacitance contribution is eliminated, and the open-gate MOS APDP would present higher impedance, which is better suited for system integration.

RF measurements on conversion gain and linearity were performed with similar frequency tones as those described earlier (RF at 1.80004 GHz, LO at 0.9 GHz, 0.45 GHz, and 0.45 GHz and 1.35 GHz combined, respectively). An open-gate NMOS APDP structure has been chosen for the following reasons. (a) It provides higher current output compared to the other APDP topologies available while having the same cut-in voltage. (b) It provides lower parasitic capacitance at the input port. Measurements were performed with a 4 dB of back-off for an open-gate APDP with 50-μm width transistors. The same back-off condition is used in order to compare the performances of GaAs MESFET and NMOS-based APDPs. The conversion loss improves by 3 dB in the case of NMOS APDPs, with a reduction in LO power level of 4dB. The measurement results are similar to the ones previously illustrated for GaAs-MESFET-based APDPs.

## 5.7. DESIGN STEPS FOR APDP-BASED RECEIVERS

In the previous sections, we have introduced the reader with different phenomena involved in subharmonic mixer design based on APDP topology. In this section, we illustrate the sequence of steps in the design of APDP-based subharmonic receivers. It is quite interesting to note at this point that the passive mixer design techniques are indeed quite different, and many nontrivial design issues need to be considered. The following are the steps taken when designing a subharmonic mixing front-end.

1. Choose the diode pair topology. This step is very much dependent on the semiconductor process technology under consideration. The choices are made from the categories described above: Schottky barrier diodes, NMOS-based diodes, and bipolar base–emitter junction diodes. Schottky diodes are the first and foremost choice due to their operation as majority carrier devices and extremely low parasitics. If they are not available, one should choose one of the other two devices.
2. Perform a DC analysis on the diode pair. This should be performed in order to verify the antisymmetric I–V characteristics of the diode pair under consideration. From the DC I–V characteristics, it is possible to estimate the peak voltage levels of the LO waveform, so that we can capture the nonlinear behavior without requiring a very high voltage drive (and hence high power consumption) from the LO buffer amplifiers. This is shown in Fig 5.19. In this figure, Vc denotes the cut-in voltage for the diodes, and Vnl is the LO amplitude required for the mixing operation to take place.
3. Choose a voltage level (amplitude), which does not correspond to one of the conversion nulls in the IF characteristics. As shown in the measured conversion loss characteristics of the APDP core, there is a voltage level at which conversion nulls are formed at IF. It is worth noting that for *N*th harmonic

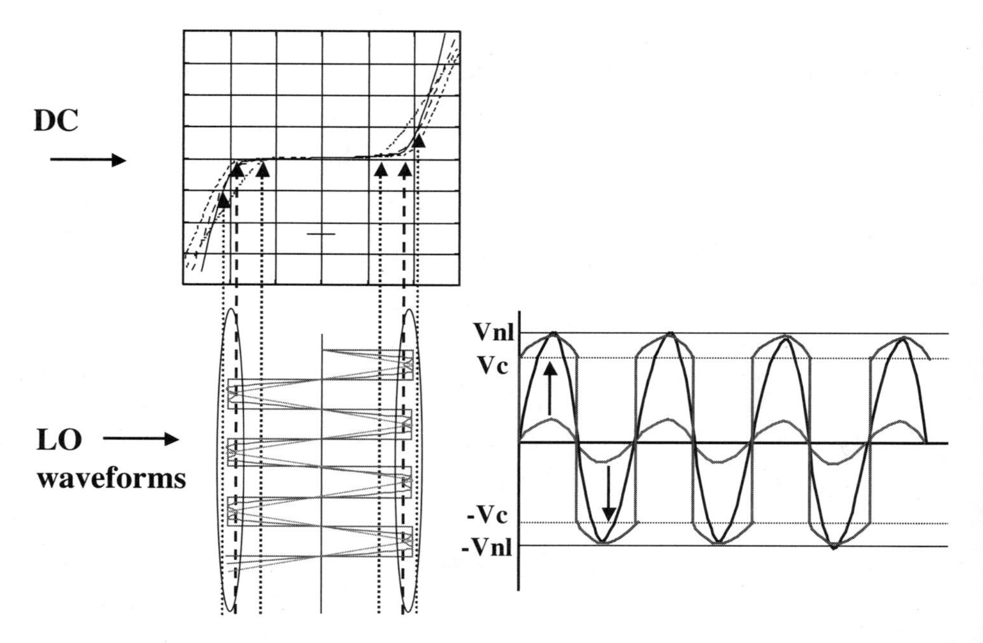


**Fig. 5.19.** Illustration of different diode transfer characteristics and corresponding waveform shapes.

operation in the case of a single-diode mixer, there are $(N-1)$ nulls in the IF characteristics. For an APDP core, the number of nulls is reduced to $(N/2-1)$. This follows from the illustration provided in [13, 14]. The details of the analysis are beyond the scope of this book, but the interested reader is encouraged to investigate the technique.

4. Choose an optimum waveform shape. A steep transition (and hence higher slope) of the waveform from zero voltage to the peak value implies more efficient switching of the diodes, and is important for performance of APDP cores with respect to conversion loss and linearity. One should also ensure that there is minimum overhead in the system implementation while generating the different waveform shapes.
5. Perform S-parameter simulation on the APDP core. This is needed for the receiver design. While performing the S-parameter simulation, the LO waveform must be present in the circuit, as in the usual operation of the APDP core. To minimize the effort, discrete samples of the time domain LO voltage waveform can be utilized as DC bias points, and the input impedance of the APDP core can be taken as an average of the individual impedance levels.
6. Design the lumped element as part of the mixer core. The lumped elements are utilized in realizing short- and open-circuited networks at RF and LO

frequencies, as appropriate. However, one concern is the design and implementation of passive components at very high frequencies in silicon substrates.

7. Choose an architecture for the mixer and receiver architectures. Differential architecture is recommended so as to minimize the effects of common mode noise.
8. Design the high-pass filter, which should be integrated at the mixer output. The cutoff frequency for this filter is a trade-off between bit-error rate and intersymbol interference.
9. Choose the LO voltage for optimum conversion loss. The mixer performance is denoted by conversion loss, linearity, and noise figure. It is quite important to note that the impedance level of the APDP changes with respect to the LO voltage. The impedance also varies with third harmonic injection technique, as described above.
10. Design the mixer for specific linearity and noise figure requirements. Three different linearity parameters are important: (a) input 1 dB compression point, (b) IIP3, and (c) IIP2. The first two are relevant for any receiver front-end, but the third is a major consideration for direct conversion receivers for amplitude-modulated interferers. We recommend the use of simulators to aid the designer in obtaining these performances. However, IIP2 should be estimated with a mismatch between the diodes. This mismatch can be taken as 5% for a pessimistic approximation. Mismatch effects should also be taken into consideration for estimating DC offset.
11. Repeat the above steps with continued optimization with respect to device dimension until the targeted specifications are met.
12. Interface mixer circuitry with other building blocks such as LNA, frequency multiplier, etc.
13. Estimate the overall receiver performance: conversion gain, IIP3, IIP2, DC offset, and input matching of the receiver.

## 5.8. ARCHITECTURAL ILLUSTRATION

Figure 5.20 shows a generic low-IF/direct conversion receiver architecture based on the proposed harmonic mixing technique. It is capable of providing separate control over the amplitude levels of different harmonics. The main advantage of this receiver architecture is the use of a very low LO frequency for down-conversion, which can be easily generated on-chip. As an example, for a 60 GHz receiver front-end, a VCO generating 7.5 GHz on-chip would be sufficient for the downconversion scheme. The in-band LO leakage would also be lower, as the incoming RF frequency is eight times higher than the applied LO frequency. However, implementing a combiner with very wide bandwidth is a challenge to deal with, and square wave generation might prove to be attractive.

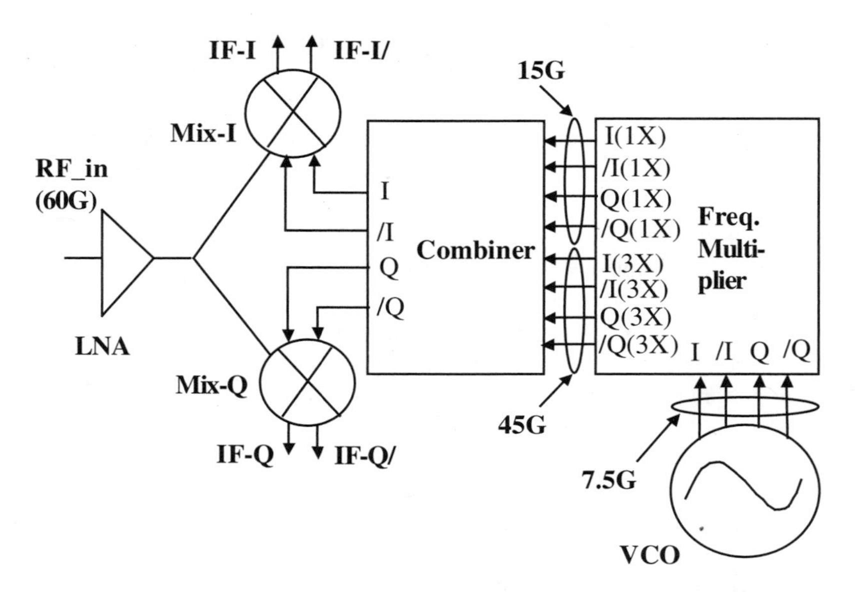


**Fig. 5.20.** A 60 GHz direct conversion receiver architecture incorporating harmonic mixing for future-generation WLAN applications.

## 5.9. FULLY MONOLITHIC RECEIVER DESIGN USING PASSIVE APDP CORES

Having introduced the basic subharmonic mixing techniques and design principles in the previous sections, we now focus on an illustration of RF front-end receiver design using passive mixing techniques with antiparallel diode pair (APDP) topology. The focus is on a direct conversion receiver design in C-band. The technology used is a 0.6 μm GaAs MESFET process.

### 5.9.1. Integrated Direct Conversion Receiver MMICs

In this section, we present two highly integrated and high-performance direct conversion receivers to address the need for high-data-rate wireless applications at 5.8 to 5.9 GHz. Figure 5.21 shows a block diagram of the receivers under consideration.

The receivers are composed of an LNA followed by a Wilkinson power divider, two even-harmonic mixers, and direct-coupled baseband amplifiers. The LO is supplied to the mixers by a modified Wilkinson power divider that incorporates a 45° phase shift in one of the LO paths. Due to the frequency-doubling action in the mixers, only 45° of phase shift is required to produce *I* and *Q* signals at the mixer outputs. The following sections illustrate the design and characterization of receiver subblocks.

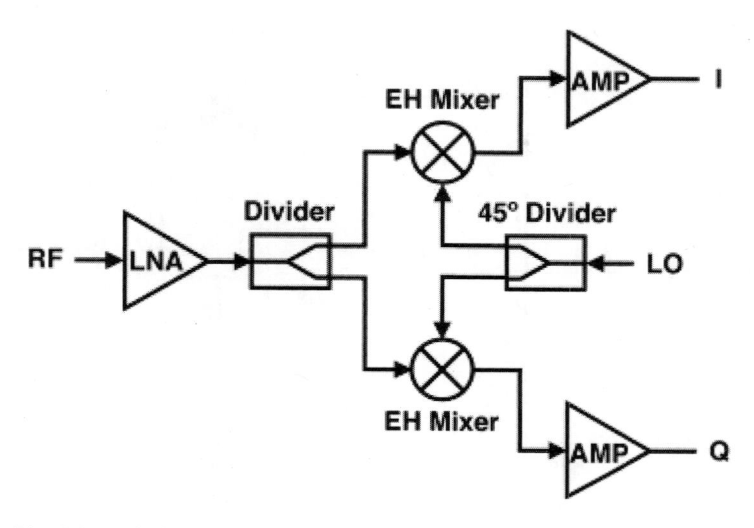


**Fig. 5.21.** Block diagram of the highly integrated direct conversion receiver.

### 5.9.2. Receiver Blocks

In this section, we illustrate two new low-power MMICs that incorporate LNAs and direct-coupled baseband amplifiers. These additions help reduce noise figure and enhance conversion gain while maintaining the required linearity and DC offset performance. Both receiver MMICs consume less than 80 mW of DC power from a 3 V supply and occupy a compact $75 \times 60$ mil$^2$ of die area.

The MMICs differ from each other in two building blocks: LNA and mixer. The high-gain receiver uses a standard LNA with subharmonic resistive FET mixers, whereas the switched-gain receiver uses a switched LNA with subharmonic APDP mixers. We will first discuss the schematics and measurement results of each block and then describe the integrated receiver MMICs.

***5.9.2.1. Low-Noise Amplifier.*** Figure 5.22 shows the circuit diagram of the LNA. The LNA is composed of a 200 μm common-source FET followed by two 300 μm FETs in a cascode configuration. Inductive degeneration in the source is used to help stabilize the LNA while moving the input noise match closer to 50 Ω impedance [3]. Enhancement-mode devices have been used to simplify DC biasing with positive supply voltages. The gate voltages are supplied through large resistors to avoid loading the gate impedance. Small resistors are also used in series in the drain line to increase immunity to wirebonds used during the measurements.

The LNA small-signal scattering parameters (S-parameters) are characterized using an HP8510 network analyzer with on-wafer ground–signal–ground (GSG) probes. The NF measurements are performed similarly with an HP8970B noise figure meter. This LNA has a gain of 21 dB, IIP3 of −10 dBm, and NF of 2.6 dB. As shown in Fig. 5.23, the LNA chip is mounted on a test board where the bias pads are

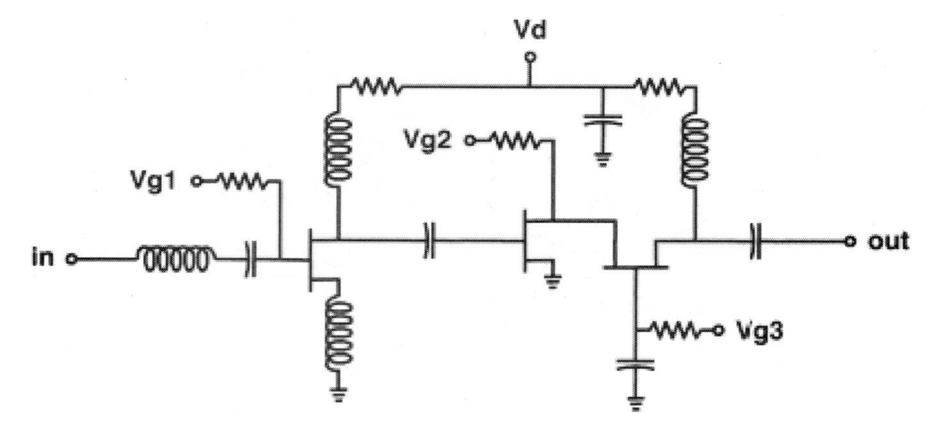


**Fig. 5.22.** Circuit schematic of the LNA.

wirebonded to the supply lines on the board. To eliminate low-frequency oscillations, all bias lines are adequately grounded on the test board by 0.1 μF and 15 pF chip capacitors. The drain voltage is biased at 3 V and all three gate–source voltages are set at 0.4 V for 8.5 mA of current from the drain supply.

***5.9.2.2. Switched-Gain LNA.*** As shown in Fig. 5.24, the switched LNA is composed of the two-stage LNA as described in the previous section connected in paral-

**Fig. 5.23.** Test board used for the LNA measurements.

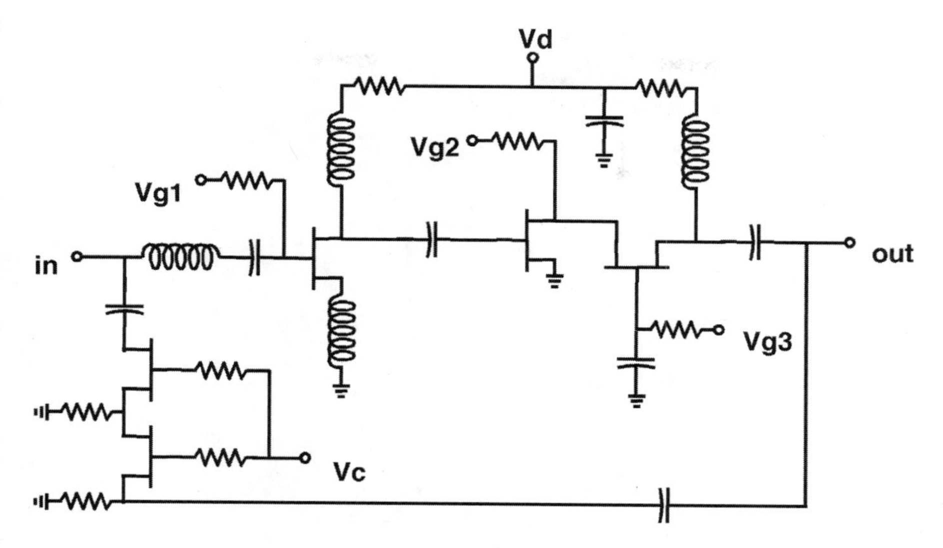


**Fig. 5.24.** Circuit schematic of the switched-gain LNA.

lel with a bypass switch [4]. The matching network in the LNA is modified to account for the loading effects of the switch and to maintain a good 50 Ω match when the LNA is turned off. In the high-gain mode, the LNA is turned on, and the switch is turned off to reduce feedback. This mode of operation provides the gain and NF required for good receiver sensitivity. In the low-gain mode, the switch is fully turned on, and the LNA is turned off to avoid the generation of intermodulations. In this mode of operation, the linearity of the LNA is improved to increase the dynamic range.

The switch is composed of two 300 μm enhancement-mode FETs connected in series. This switch provides an IIP3 of +17 dBm, and loss of 7 dB with negligible DC current. Two FETs are used in the switch to increase the reverse isolation. The gain of the switched LNA is reduced in the high-gain mode because of the capacitive feedback of the switch.

The measurements of the switched LNA are performed in a similar setup as the one used for the LNA. Figure 5.25 shows the test board used for the measurements of the switched LNA. For the high-gain mode, the LNA drain voltage is biased at 3 V with a drain current of 16 mA while the switch FETs are turned off by grounding the gate voltages.

In the high-gain mode, the switched LNA shows a gain of 11 dB, IIP3 of –5 dBm, and NF of 3.5 dB. A shunt off-chip inductor of 1 nH is used to improve the input return loss to 15 dB with a corresponding output return loss of 8 dB at 5.85 GHz. The noise figure of the switched LNA is considerably increased due to additional switch noise and its impact on the shift in the optimum noise-matching point. Better modeling of the switch and its noise parameters can help to predict such performance degradations in the design. For the low-gain mode, the LNA is turned off by grounding the gates while the switch is turned on by setting the gate voltages on

**Fig. 5.25.** Test board used for the switched LNA measurements.

the switch FETs to 3 V. The switched LNA maintains better than 10 dB of input and output return loss with 7 dB of loss in the low-gain mode. Our measurements also show an average NF of 7.5 dB and IIP3 of 17 dBm for negligible DC current.

***5.9.2.3. Subharmonic APDP Mixer.*** Each mixer is composed of a pair of 150 μm Schottky-barrier diodes in antiparallel diode pair (APDP) configuration, interconnected with resonant tanks, that are used for port-to-port isolation. To minimize the impact of diode mismatch on DC offset and IM2 performance of EH mixers, interdigitated APDP structures are used to improve matching and reduce such unwanted effects. Resonant tanks are made up of metal–insulator–metal (MIM) capacitors connected in parallel with spiral inductors. One resonant tank is used to isolate both the RF and IF ports from the LO signal, whereas the other two are used to isolate the IF port from both the RF and LO signals. By using APDP mixers, we are able to increase IIP2, alleviate in-band LO leakage, and reduce the time-variant DC offset. Because of the frequency scheme of such mixers, LO frequency of 2.9 GHz is used to down-convert the RF band of interest at 5.9 GHz. Figure 5.26 shows the schematic of the subharmonic APDP mixers.

Measurements were performed on-wafer using GSG coplanar probes. The mixers exhibit a voltage conversion loss of 11 dB, BW of 2.5 MHz, IIP3 of +23 dBm, and IIP2 of above +51 dBm, with an LO drive level of +13 dBm.

***5.9.2.4. Subharmonic Resistive FET Mixer.*** In this mixer design, a novel topology for a resistive FET mixer is developed to improve isolation by taking ad-

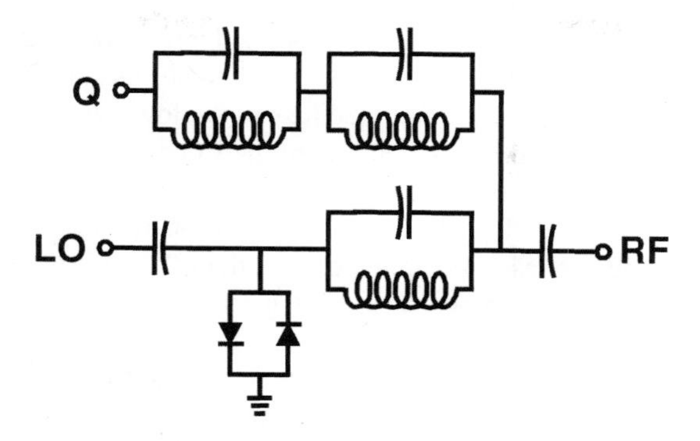


**Fig. 5.26.** Circuit schematic of the subharmonic APDP mixer.

vantage of a three port mixing device instead of a two port diode. The schematic of the subharmonic resistive FET mixer is shown in Fig. 5.27.

Each mixer is composed of a pair of 600 μm resistive FETs that is pumped by the LO signal at the gate through a spiral coupled-line transformer. This type of transformer provides great performance with small size implementation [5]. A resonant tank made up of a metal–insulator–metal (MIM) capacitor and a spiral inductor is used in this mixer to isolate the RF and IF ports at the drain of the FETs. Mixers show a voltage conversion loss of 14 dB, BW of 2.6 MHz, IIP3 of +8 dBm, and IIP2 of above +6 dBm, with a LO drive level of +10 dBm. These measurement results deviate from those predicted by simulations, which can be attributed to poor transformer design and large mismatch between the FETs. High loss in the transformer forces the mixers into starved LO operation, resulting in poor conversion characteristics.

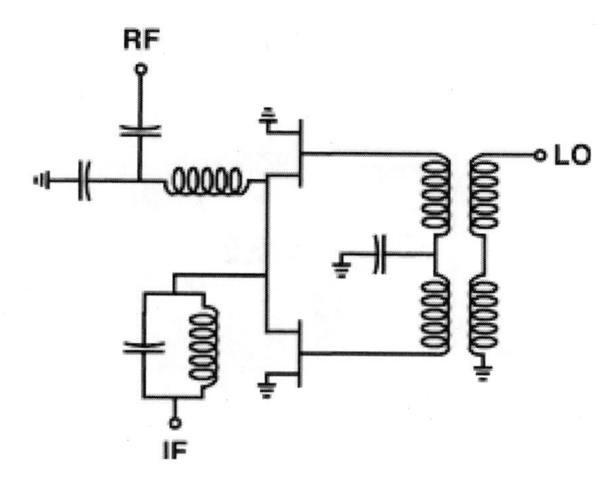


**Fig. 5.27.** Circuit schematic of the subharmonic resistive FET mixer.

***5.9.2.5. LO Divider.*** The LO divider is designed with a Wilkinson power divider topology with a 45° low-pass structure on one of the outputs and a series resistor on the other. The resistor is used to match the loss in the low-pass structure and reduce amplitude imbalance. The circuit schematic of the LO divider is shown in Fig. 5.28.

The response of the divider is measured using on-wafer probes. It exhibits a loss of 4.5 dB at 2.9 GHz with 10 dB of input return loss. An amplitude imbalance of less than 0.5 dB and a phase imbalance of less than 0.3° are also measured at 2.9 GHz. These measurements are performed using 50 Ω terminations on the outputs of the divider. In practice, these ports are terminated by the mixers, thus a 50 Ω match is very unlikely. Although the identical impedance at the LO port of the two mixers helps to maintain the relative phase and amplitude imbalance, the shift from 50 Ω impedance at the mixer–divider interface reduces the accuracy of the measurements. To alleviate this possible inaccuracy, the phase and amplitude imbalance need to be remeasured in real time by observing the output $I$ and $Q$ signals of the receiver. This is further explained in the next section.

***5.9.2.6. Integrated High-Gain Receiver.*** Having illustrated the different building blocks, we now describe the integrated high-gain receiver. The circuit schematic of the high-gain receiver is shown in Fig. 5.29. This receiver consists of a LNA, a Wilkinson RF divider, two subharmonic resistive FET EH mixers, and direct-coupled baseband amplifiers.

This receiver chip is mounted on a test board and the bias pads are wirebonded to the board for proper AC grounding using bypass capacitors. The RF and LO signals are applied on-wafer by GSG probes, whereas the outputs are taken off-chip to the board through wirebonds. Vias are used throughout the board to ensure good AC grounding. Figure 5.30 shows a picture of the test board used for the measurements. A summary of the receiver measurements is shown in Table 5.1.

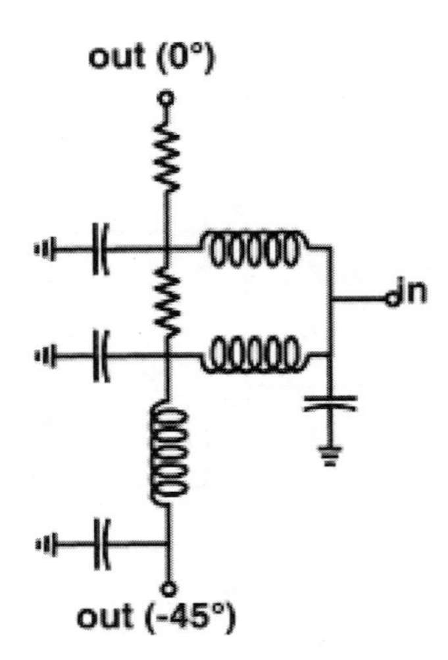


**Fig. 5.28.** Circuit schematic of the LO divider.

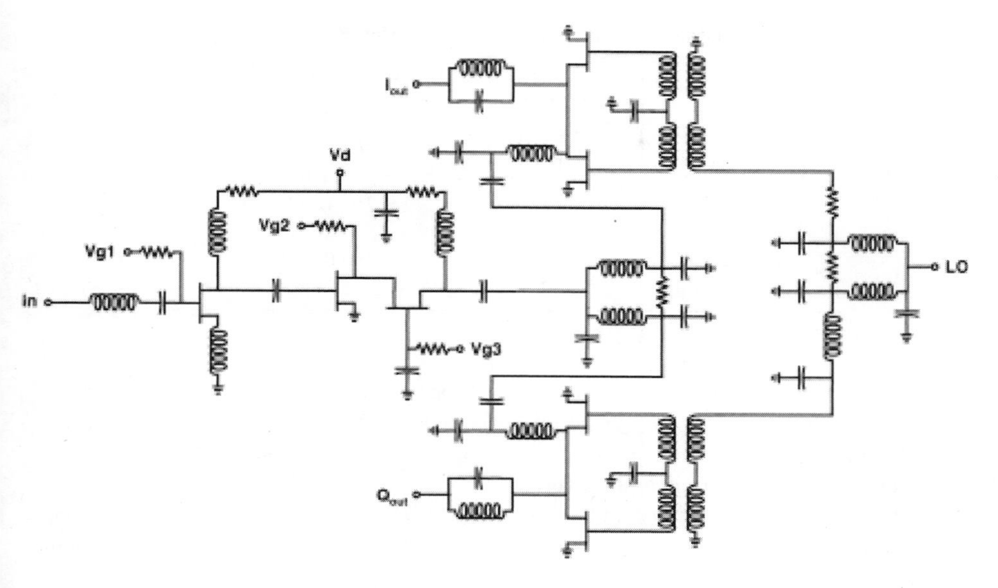


**Fig. 5.29.** Circuit schematic of the high-gain receiver MMIC.

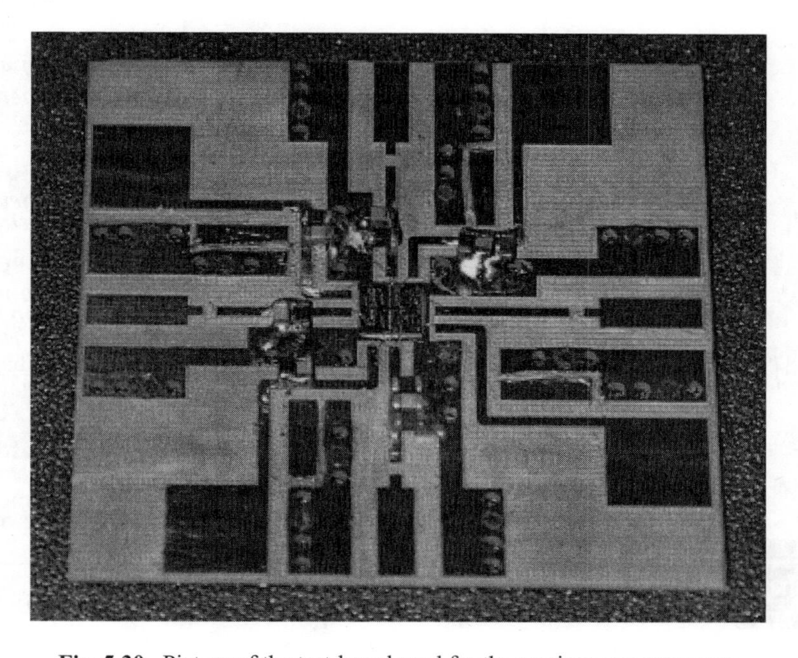

**Fig. 5.30.** Picture of the test-board used for the receiver measurements.

**Table 5.1.** Summary of the high-gain receiver measurement results

| Performance | Measured result |
|---|---|
| Gain | 4 dB |
| NF | 4.6 dB |
| IIP3 | –10 dBm |
| IIP2 | 10 dBm |
| LO Leakage | < –75 dBm |
| *I*/*Q* Imbalance | 4° and 0.3 dB |
| Power consumption | 31.5 mW |

As is evident from Table 5.1, the overall receiver results suffer from the performance of the mixer and baseband amplifiers. The LNA and the LO phase shifter show reasonable performance, and LO leakage is easily kept under specification by virtue of the subharmonic architecture.

The phase and amplitude imbalance need to be remeasured in real time by observing the output *I* and *Q* signals of the receiver. This measurement is performed with the high-gain receiver in the test setup shown in Fig. 5.30. Figure 5.31 shows

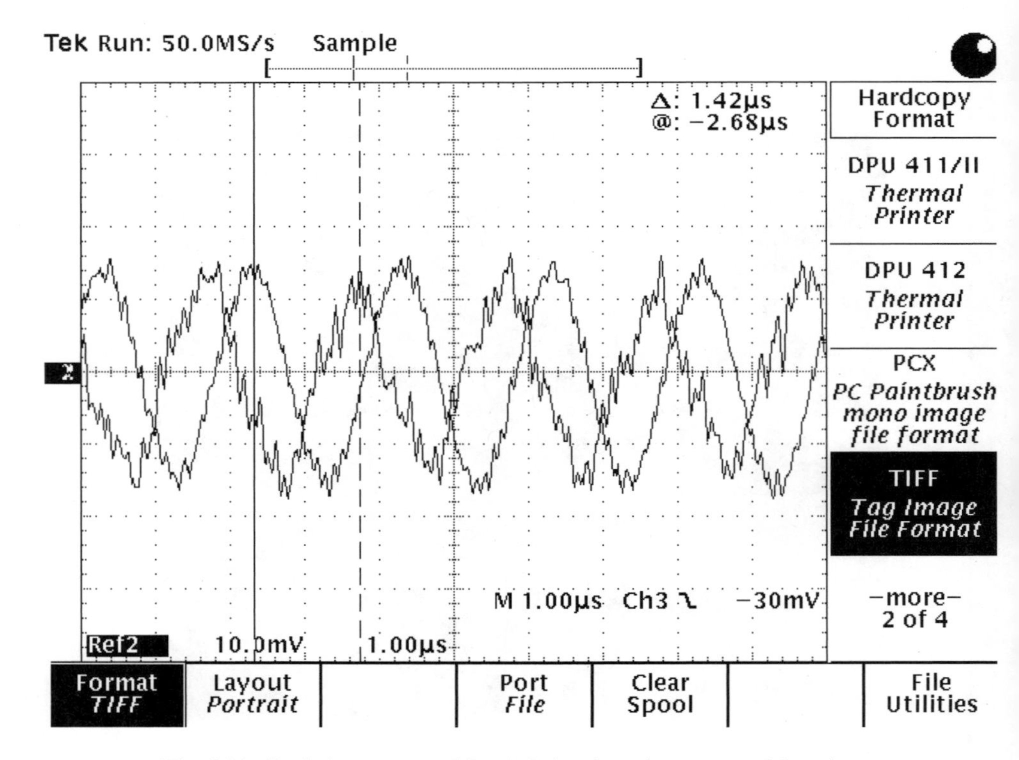


**Fig. 5.31.** Real-time measured *I* and *Q* signals at the outputs of the mixers.

the real-time 51 kHz $I$ and $Q$ outputs of the mixers. The measured time difference of 1.42 μs translates to 94° of phase difference between the two channels. This is slightly larger than the predicted phase imbalance of less than 1°.

***5.9.2.7. Integrated Switched-Gain Receiver.*** The circuit schematic of the switched-gain receiver is shown in Fig. 5.32. This receiver integrates a switched LNA, a Wilkinson RF divider, two subharmonic APDP mixers, and direct-coupled baseband amplifiers.

This receiver chip is tested similarly to the high-gain receiver and a summary of the receiver measurements is shown in Table 5.2. Although the noise figure measured below the specified 6 dB, considering the lack of gain in the front-end, that leaves little margin for the noise that would have been contributed by the baseband amplifier. However, the receiver shows excellent linearity in both high- and low-gain modes and consumes little DC power. With additional gain in the switched LNA, this receiver can be a good candidate for most applications requiring high dynamic range.

### 5.9.3. Additional Receiver Blocks

Several other receiver blocks have been designed and developed but not integrated with the receiver MMICs described in the previous sections. Additional integration would have increased the die size to a limit beyond which yield of the ICs can suffer in high-volume production. These blocks are fabricated and tested on separate chips.

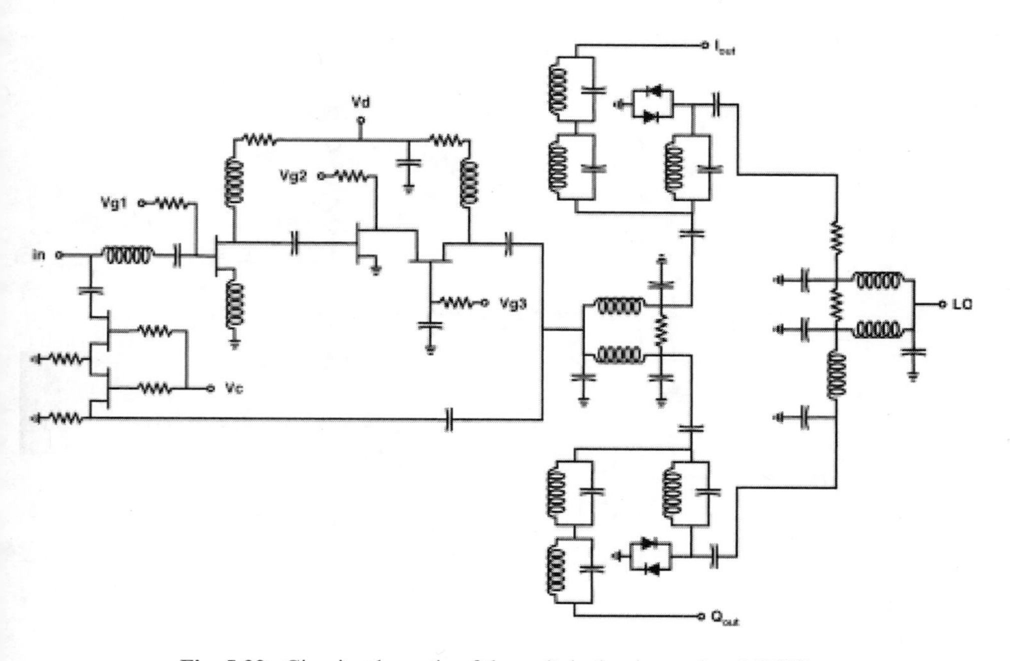


**Fig. 5.32.** Circuit schematic of the switched-gain receiver MMIC.

**Table 5.2.** Summary of the switched-gain receiver measurement results

| Performances | MMIC in high-gain mode | Measured MMIC in low-gain mode |
|---|---|---|
| Gain | –3 dB | –22 dB |
| NF | 5.6 dB | 23 dB |
| IIP3 | –5 dBm | 17 dBm |
| IIP2 | 44 dBm | 62 dBm |
| LO Leakage | < –70 dBm | < –70 dBm |
| *I*/*Q* Imbalance | 4° and 0.3 dB | 4° and 0.3 dB |
| Power consumption | 48 mW | <2 mW |

***5.9.3.1. LO Buffer Amplifier.*** The schematic of the LO buffer amplifier is shown in Fig. 5.33. The LO buffer is a two-stage common-source amplifier with 600 μm FET in the first and an 800 μm FET in the second stage. Interstage matching is provided by the high-pass network at the drain of the first FET.

The amplifier is mounted on a test board similar to the one described earlier for proper AC grounding and biased at a drain current of 38 mA from a 3 V supply. Small-signal S-parameter measurements are performed using an HP8510 network analyzer. The buffer amplifier shows a gain of 24 dB and an output $P_{1dB}$ of +10 dBm.

***5.9.3.2. Transimpedance Baseband Amplifier.*** The baseband amplifier is designed as a transimpedance amplifier to provide a match between the low impedance of a diode mixer and the high- input impedance of a baseband amplifier. The

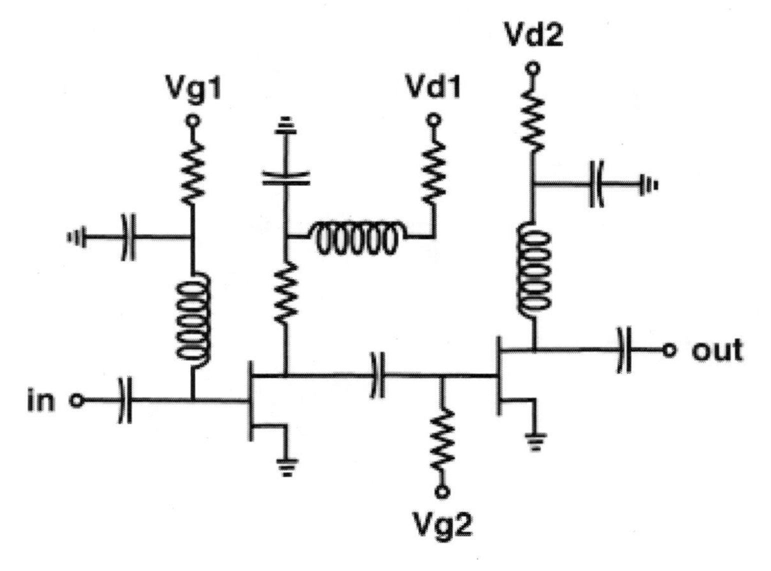


**Fig. 5.33.** Circuit schematic of the LO buffer amplifier.

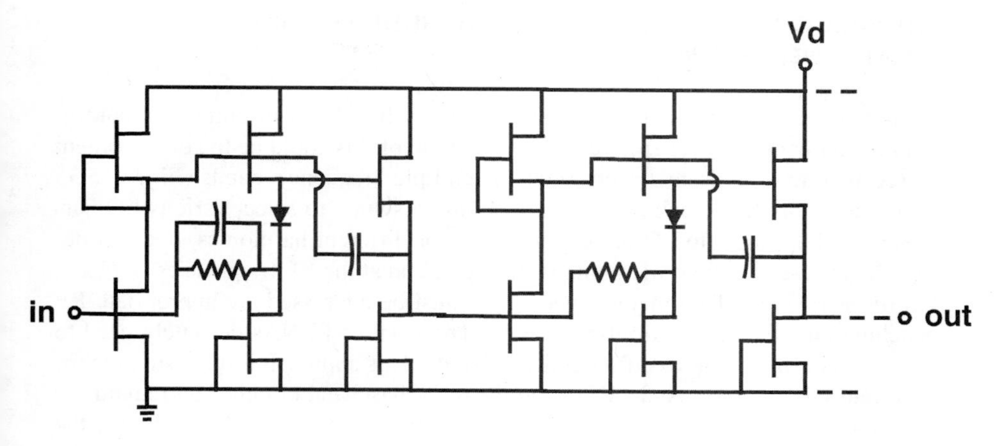


**Fig. 5.34.** Circuit schematic of the transimpedance baseband amplifier.

circuit schematic of this amplifier is shown in Fig. 5.34. This amplifier is biased at 6 mA of DC current from a 3 V supply. The amplifier shows gain of 20 dB with 3 dB BW of 48 MHz.

A picture of the complete die is shown in Fig. 5.35. Both receivers and the test structures used for characterization of the individual blocks are implemented in this die. It was fabricated with the TriQuint TQTRx GaAs MESFET process.

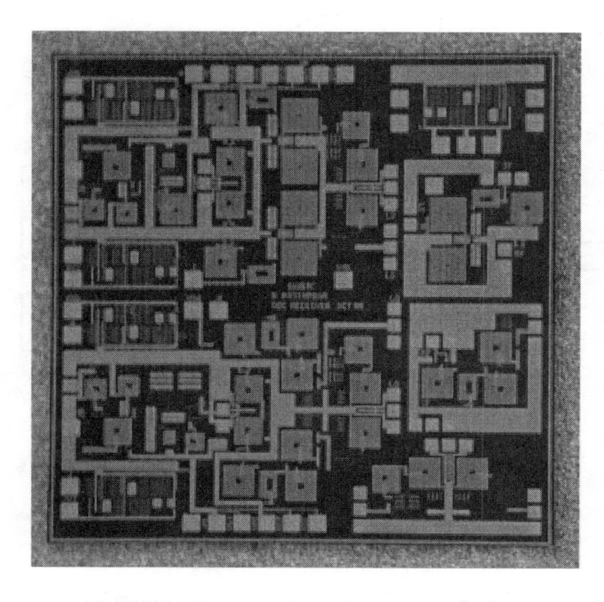

**Fig. 5.35.** The complete 120 × 120 $mil^2$ die.

## 5.10. RECONFIGURABLE, MULTIBAND SUBHARMONIC FRONT-ENDS

Subharmonic receiver architectures are extremely beneficial in building reconfigurable multiband receiver front-ends. An example of this could be the development of receiver front-ends to be operated in multiple frequency bands. Figure 5.36 shows an example of such architectures. Using a switch to select various LO harmonics and appropriate LO amplitude levels for different harmonics, one can design the front-end to be programmable for operation at the RF frequencies of $F$, $2F$, $4F$, $6F$, and so on. Two major aspects that must be addressed are appropriate RF matching circuits and the front-end LNAs. The front-end LNAs definitely need to be custom designed for specific frequency bands. This approach is quite suitable for most reusable designs, as the VCO and the other baseband circuits can remain the same for each subharmonic operation, thus minimizing the effort of redesigning the front-end as processes and technologies scale. With wideband operation of the front-end, this could become a potential candidate for covering the major WLAN bands of use. However, one drawback of these architectures is the multiplication of LO phases by integers along with the harmonic order. For this reason, all the phase shifts are usually avoided in the LO path and incorporated in the RF path. For example, if we want to provide phase shifts of 0° and 180°, respectively, in fourth-harmonic operation, the VCO needs to generate 45° signals. One such approach is described with detailed illustrations in Chapter 6. This effect becomes more signifi-

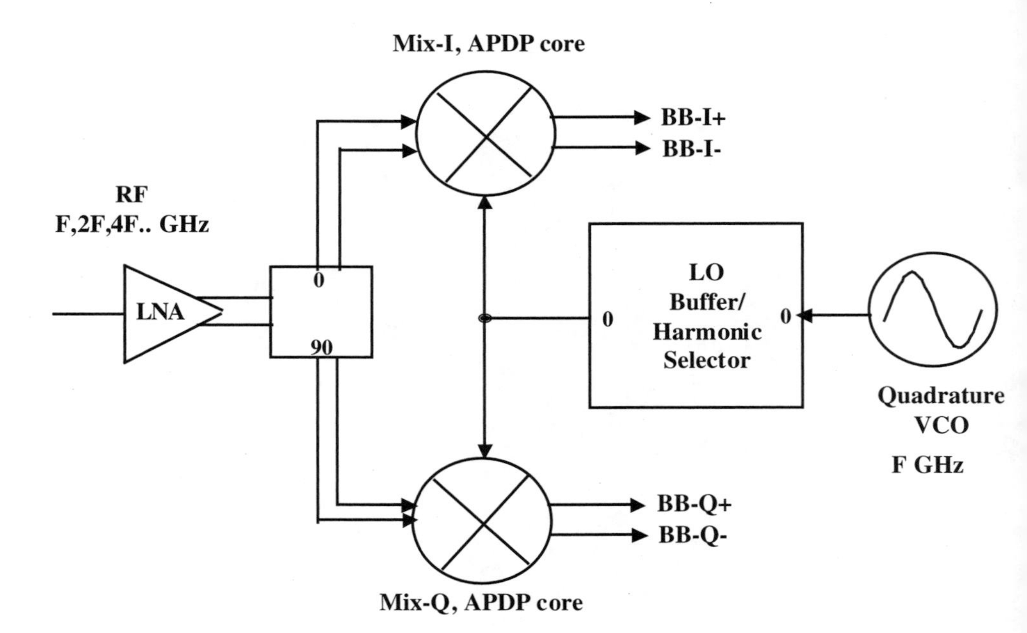


**Fig. 5.36.** Reconfigurable, multiband RF front-end using subharmonic approach.

cant in the case of higher-order subharmonics and, thus, is quite suitable for providing the phase shift in the RF path.

## 5.11. CONCLUSION

In this chapter, we have described the theory and design of subharmonic mixers, leading to the development of a fully monolithic receiver. The nonlinear characteristics of the mixer cores have been explored, along with additional LO harmonic injection. An extension of this approach is shown as a potential candidate for multiband architecture. Subharmonic-mixer-based receiver architectures are quite capable of being a low-power solution in multigigahertz receivers. In the subsequent chapters, we will focus on other approaches for subharmonic receiver design.

## REFERENCES


1. B. L. Sharma, *Metal–Semiconductor Schottky Barrier Junctions and Their Applications,* Wiley, 2002.
2. M. Cohn, J. E. Degenford, and B. Newman, "Harmonic mixing with an antiparallel diode pair," *IEEE Transactions on Microwave Theory and Techniques, MTT-23,* 667–673, August 1975.
3. S. Yoo, M. R. Murty, D. Heo, and J. Laskar, "A C-band low power high dynamic range GaAs MESFET low noise amplifier," *Microwave Journal, 43,* 2, 90–106, Feb. 2000.
4. H. Morkner, M. Frank, and S. Yajima, "A miniature PHEMT switched-LNA for 800 MHz to 8.0 GHz handset applications," in *IEEE RFIC Symposium,* pp. 109–112, June 1999.
5. C. Trantanella, M. Shifrin, and B. Bedard, "Low cost, plastic encapsulated mixers for C/X-band applications," in *GaAs IC Symposium,* pp. 131–134, Nov. 1998.
6. G. W. Neudeck, *The PN Junction Diode,* Addison-Wesley, 1989.
7. B. Razavi, *Monolithic Phase-Locked Loops and Clock Recovery Circuits Theory and Design,* IEEE Press, 1996.
8. M. Shimozawa, K. Kawakami, H. Ikematsu, K. Itoh, N. Kasai, Y. Isota, and O. Ishida, "A monolithic even harmonic quadrature mixer using a balanced type 90 degree phase shifter for direct conversion receivers," in *IEEE Microwave Theory and Technique Symposium,* pp. 175–178, June 1998.
9. B. Matinpour, S. Chakraborty, and J. Laskar, "Novel DC-offset cancellation techniques for even-harmonic direct conversion receivers," *IEEE Trans. Microwave Theory and Tech., 48,* 12, 2554–2559, December 2000.
10. W. Zhao, C. Schollhorn, E. Kasper, and C. Rheinfelder, "38GHz coplanar harmonic mixer on silicon," in *Silicon MonolithicIntegrated Circuits in RF Systems,* pp. 138–141, 2001.
11. B. K. Kormanyos and G. M. Rebeiz, "A 30-180 GHz harmonic mixer-receiver," in *IEEE International Microwave Symposium,* vol. 1, pp. 341–344, 1992.

12. M. Shimozawa, T. Katsura, N. Suematsu, K. Itoh, Y. Isota, and O. Ishida, "A passive-type even harmonic quadrature mixer using simple filter configuration for direct conversion receiver," in *IEEE International Microwave Symposium,* pp. 517–520, June 2000.
13. J. Hesler, D. Kurtz, and R. Feinaugle, "The cause of conversion nulls for single-diode harmonic mixers," *IEEE Microwave and Guided Wave Letters, 9,* 12, 532–534, December 1999.
14. R. Feinaugle, H.-W. Hubers, H. P. Roser, and J. Hesler, "On the effect of IF power nulls in Schottky diode harmonic mixers," *IEEE Transactions on Microwave Theory and Techniques, 50,* 1, 134–142, January 2002.
15. D. N. Held and A. Kerr, *Conversion Loss and Noise of Microwave and Milimeter-Wave Mixers. Part 1—Theory,"* Plenum Press, 1984.
16. H. C. Torrey and C. A. Whitmer, *Crystal Rectifiers,* MIT Radiation Lab. Series, Vol. 15, McGraw-Hill, 1948.
17. Y. Tao, Y. X. Bo, L. R. Hou, W. Tiehua, and Z. Monglin, "Ka-band preselected harmonic mixer," in *International Conference on Infrared and Milimeter Waves,* pp. 455–456, 2000.
18. K. Itoh, A. Iida, Y. Sasaki, and S. Urasaki, "A 40GHz band monolothic even harmonic mixer with an antiparallel diode pair," *IEEE MTT-S Digest,* pp. 879–881, 1991.
19. L. Sheng, J. C. Jensen, and L. Larson, "A Wide-bandwidth Si/SiGe HBT direct conversion sub-harmonic mixer/downconverter," *IEEE Journal of Solid-State Circuits, 35,* 9, 1329–1337, September 2000.
20. J.-D. Buchs and G. Begemann, "Frequency conversion using harmonic mixers with resistive diodes," *IEEE Journal on Microwave, Opt. Accoustics Theory, 2* 71–76, May 1978.
21. P. Henry, B. Glance, and M. Schneider, "Local-oscillator noise cancellation in the subharmonically pumped down-converter," *IEEE Trans. Microwave Theory and Tech., MTT-24,* 254–257, May 1976.

# 6

# ACTIVE SUBHARMONIC RECEIVER DESIGNS

## INTRODUCTION

In the previous chapter, we presented a detailed illustration of the subharmonic passive mixing technique and its applicability for practical receiver designs. However, an active subharmonic mixer-based receiver is quite attractive from a system perspective, in terms of system budget and the requirement of low drive on the LO signal waveform. An alert reader can easily see that the active subharmonic techniques can be philosophically thought of as a combination of the material introduced in the previous two chapters. Chapter 5 explored subharmonic mixing techniques using exponential characteristics of the diode pair. In this chapter, we focus on subharmonic techniques based on frequency doubling methodology. Active subharmonic mixers are becoming more and more popular, and have been reported in recent literature [1, 2, 3]. In this chapter, we cover the basic theory of operation of the active subharmonic mixers, with an illustration of their integration in the RF front-end architectures. Along with the active subharmonic mixing techniques, wideband performance of the receiver front-end becomes very much essential. This chapter first introduces the subharmonic mixing technique, followed by considerations of wideband operation.

Almost all of the active mixers reported in the literature to date [1, 2, 3] use some sort of doubler topology as a part of their core. The doubler topology can be implemented in a standard Gilbert cell core [4] fashion, or in a parallel transistor pair.

*Modern Receiver Front-Ends.* By J. Laskar, B. Matinpour, and S. Chakraborty
ISBN 0-471-22591-6 

## 6.1. STACKING OF SWITCHING CORES

### 6.1.1. Description and Principles

Figure 6.1 shows the simplified schematic of a subharmonic mixer utilizing two stages of a Gilbert switching cell. Both stages are driven by an LO signal whose frequency is half the incoming RF signal. The output voltage, then can be expressed as

$$V_{\mathrm{IF}} = V_{\mathrm{RF}}[V_{\mathrm{LOI}} \cdot V_{\mathrm{LOQ}}] \tag{6.1}$$

where all of the above voltages can be expressed in the time domain. The waveforms $V_{\mathrm{LOI}}$ and $V_{\mathrm{LOQ}}$ are at the same frequency (half of the incoming RF signal) and 90° out of phase. The Gilbert cell switching core generates the doubled frequency waveform, which mixes with the incoming RF frequency, and then is down-converted to the baseband.

One potential problem associated with this topology is the larger supply voltage that is required. To address this issue, the RF section can be biased in parallel with, and coupled to, the LO switching section. This configuration is shown in Fig 6.2. The power required for the local oscillator is approximately double that required for a standard double-balanced mixer. Also, the noise distortion contribution from the upper switching pair is increased, but to a lesser extent.

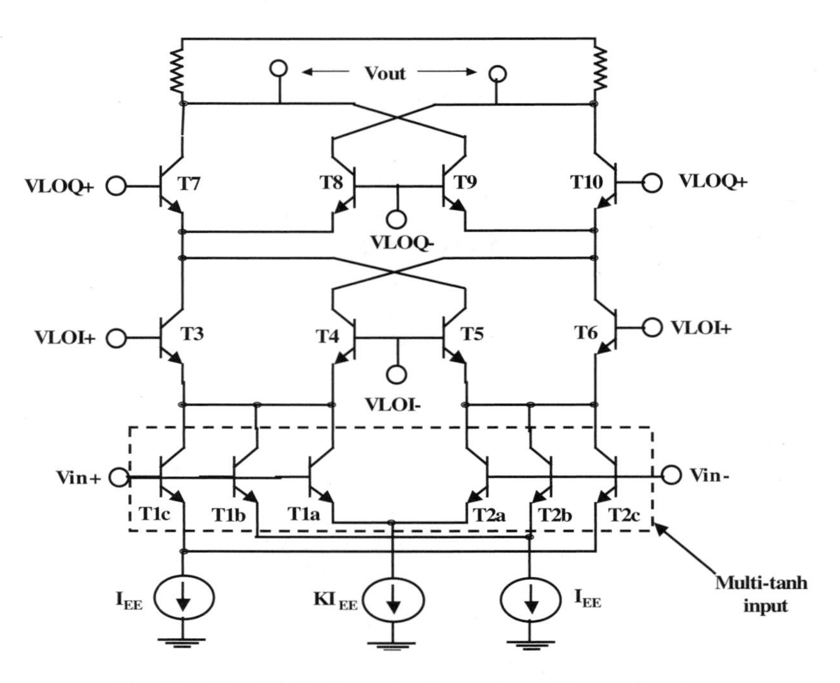


**Fig. 6.1.** Simplified schematic of an active subharmonic mixer.

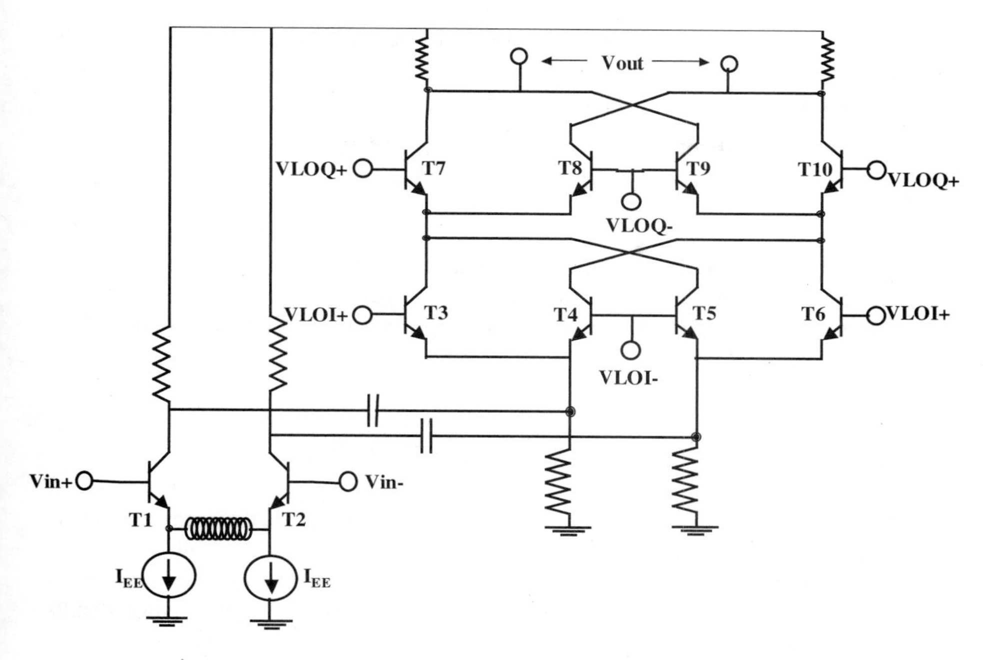


**Fig. 6.2.** Simplified schematic of an active subharmonic mixer for low voltage.

LO rejection is another improvement for the doubler-based subharmonic mixer over 1× direct conversion receivers. For a 1× direct conversion receiver, the LO signal and the RF signals share the same band, and the LO rejection is 0 dB, as described in [1]. For a doubler-based approach, imperfections in the circuit and signal matching can cause the LO signal to be down-converted to DC. If the matching between transistors and signal phases ($V_{\mathrm{LOI}}$ and $V_{\mathrm{LOQ}}$) were perfect, then no LO signal would be down-converted. Presence of phase error between the LO signals causes an LO signal at the RF input port to be down-converted to DC. A quantitative analysis of phase error and duty cycle error for these mixers is presented in [1]. Assuming that the LO rejection is defined as

$$LOR = \frac{G(\omega_{\mathrm{RF}})}{G(\omega_{\mathrm{LO}})} \tag{6.2}$$

where $G(\omega_{\mathrm{RF}})$ and $G(\omega_{\mathrm{LO}})$ are the conversion gains at RF and LO frequencies, respectively, it is shown in [1] that the LO rejection is mostly a function of duty cycle error and quite insensitive to the phase difference error between the two LO signals in quadrature.

To increase the voltage handling capability of a differential pair, a multi-tanh differential pair topology can be used. Figure 6.1 shows the input multi-tanh circuitry along with the mixer switching core. In the multi-tanh linearized differential pair, the junction areas can be optimized to obtain the best dynamic range. Assuming the bias current in the center transistors are optimized to be $K$ times the outer bias currents, the transconductance of a multi-tanh input is given by

$$Gm_0 = Gm(0) = \left(\frac{I_{\mathrm{EE}}}{2V_{\mathrm{T}}}\right)\left(\frac{K(1+A)^2 + 8A}{(1+A)^2}\right) \tag{6.3}$$

With a current scaling factor of 13, the voltage capacity of a multi-tanh circuit improves by a factor of 12 from a simple differential pair.

From a noise perspective, the total input-referred noise of the multi-tanh differential pair can be given by

$$V_{\mathrm{noise}} = 2V_{\mathrm{T}}\sqrt{\frac{2q}{I_{\mathrm{EE}}}}\left(\frac{1+A}{\sqrt{K(1+A)^2 + 8A}}\right) \tag{6.4}$$

Another important parameter for direct conversion mixers is the second-order intercept point (IIP2). This requirement is due to the presence of the high level of interferers along with the received signal. When down-converted, the unwanted interferers fall in-band along with the message spectrum. A much worse situation results when the interferer is modulated. This leads to the degradation of the signal-to-noise ratio (SNR) of the receiver.

From a circuit perspective, the second-order intercept point is the result of second-order distortion terms generated in the circuit. An analysis of the second-order products is presented in [1]. The input differential pair generates second-order distortion through the finite impedance of the current source and mismatch between the devices. The second-order current is comprised of two components: (1) the common-mode component, and (2) the differential-mode component. The three major sources of second-order distortion in a double-balanced mixer can be attributed to (a) resistor mismatch in the load, (b) duty cycle mismatch in the switching core, (c) overall duty cycle errors. A mathematical formulation of IIP2 can be given by

$$IIP_2 \cong \left\{2\left[\frac{1}{V_{\mathrm{A}}}\left(\left|\frac{\Delta R}{R}\right| + \left|\frac{\Delta d}{d}\right|\right) + \frac{2(1-2d)}{V_{\mathrm{T}}}\left(\left|\frac{\Delta I_{\mathrm{Q}}}{I}\right|\right)\right]\right\}^{-1} \tag{6.5}$$

where, $V_{\mathrm{A}}$ is the Early voltage, $\Delta R$ is the resistor mismatch, $\Delta d$ is the duty cycle mismatch between the left and right branches, and $\Delta I_{\mathrm{Q}}$ is the difference in the DC bias of the emitter currents of the two input transistors.

The above mixer architecture also can be realized in a CMOS-based technology as well. In the case of CMOS-based technology, the switching core transistors are replaced by NMOS transistors, as are the input differential pair transconductance transistors. However, the principles of operation are exactly the same as the ones described above in the case of the bipolar-transistor-based core, in terms of frequency doubling and mixing.

### 6.1.2. Subharmonic Receiver Architecture

Figure 6.3 shows a subharmonic receiver architecture that is quite favorable for the subharmonic mixers under consideration. This includes a broadband phase shifter, necessary buffers, and the subharmonic mixer cores. The mixers are driven in half-quadrature with respect to each other. Figure 6.4 shows the circuit schematic of the broadband phase shifter.

With the following choice of parameters in the phase shifter circuitry,

$$\begin{aligned} R_1 &= 8R_2,\ R_4 = 8R_3 \\ R_1C_1 &= R_2C_2 = 1/(2\pi f_1) \\ R_3C_3 &= R_4C_4 = 1/(2\pi f_2) \end{aligned} \tag{6.6}$$

the amplitudes can be given as

$$|V_\mathrm{I}| = |V_\mathrm{Q}| = \frac{1}{2}V_\mathrm{in} \tag{6.7}$$

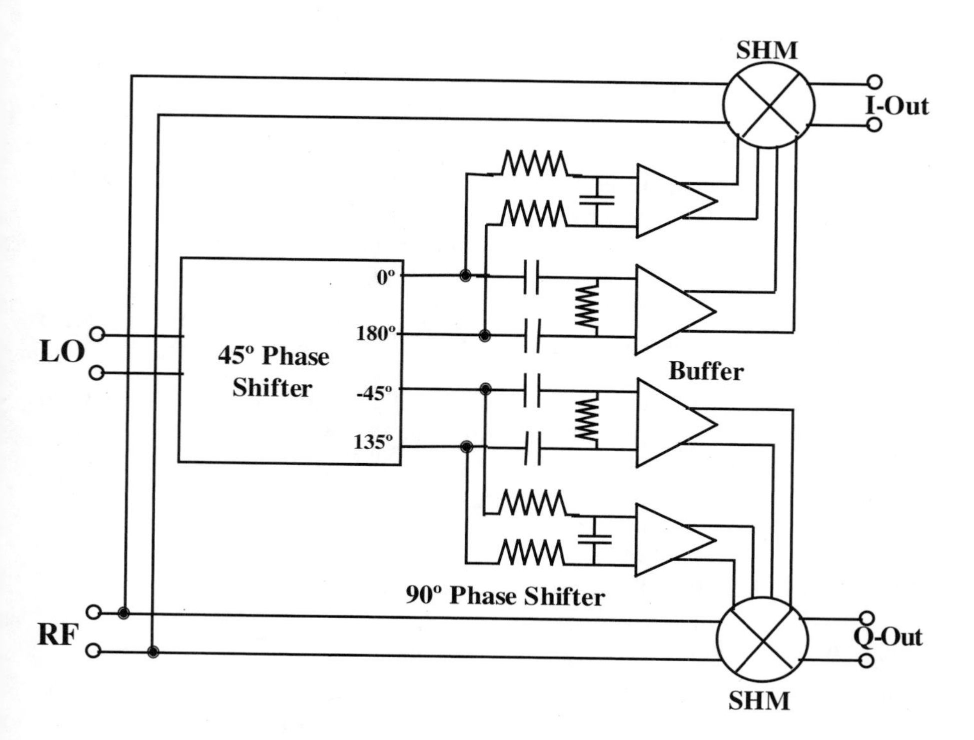


**Fig. 6.3.** Subharmonic receiver architecture.

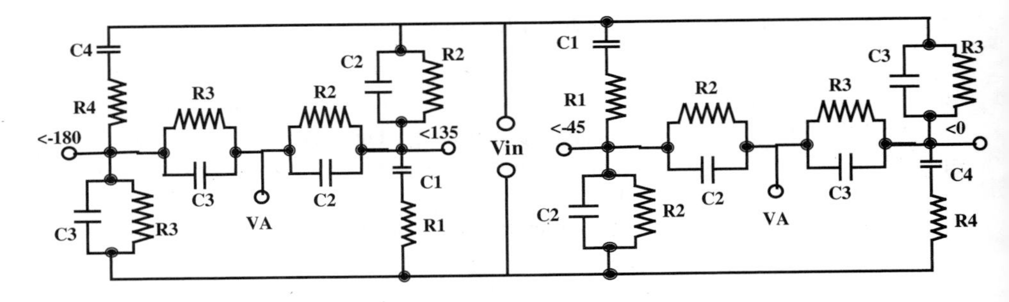


**Fig. 6.4.** Schematic of a broadband phase shifter.

The differential in-phase and quadrature voltages can be formulated as

$$
\begin{aligned}
V_{\mathrm{I}} &= V_{\text{out} < 0} - V_{\text{out} < 180} \\
V_{\mathrm{Q}} &= V_{\text{out} < -45} - V_{\text{out} < 135}
\end{aligned}
\tag{6.8}
$$

The phases of the output voltages can be given by

$$
\begin{aligned}
\phi_{\mathrm{Q}} &= \tan^{-1}\left(\frac{12F_1(1-F_1^2)}{F_1^4 - 38F_1^2 + 1}\right) \\
\phi_{\mathrm{Q}} &= \tan^{-1}\left(\frac{12F_2(1-F_2^2)}{F_2^4 - 38F_2^2 + 1}\right)
\end{aligned}
\tag{6.9}
$$

where $F_1$ and $F_2$ are normalized frequencies, denoted by, $F_1 = f/f_1$, and $F_2 = f/f_2$. The poles, $f_1$ and $f_2$ should be chosen for operation in the desired frequency range.

However, the phase difference change at the frequency $f$ is given by

$$
\Delta\phi(f) = \left[\frac{12f_1(f^2+f_1^2)}{(f^4+34f^2f_1^2+f_1^4)} - \frac{12f_2(f^2+f_2^2)}{(f^4+34f^2f_2^2+f_2^4)}\right]\left(\frac{|\Delta R|}{R} + \frac{|\Delta C|}{C}\right) \tag{6.10}
$$

This phase shifter is followed by a 90° phase shifter to generate the LO voltage to generate quadrature LO waveforms that drive the LO switching core in the subharmonic mixer. The 90° phase shifters provide additional shunt impedance to the original 45° phase shifter.

## 6.2. PARALLEL TRANSISTOR STACKS

### 6.2.1. Active Mixer

The subharmonic mixing approach described in the previous section utilizes the doubler characteristics for the large-signal LO using a doubly stacked Gilbert cell switching core. This implementation requires three levels of transistors and a quadrature LO. Stacking of transistors becomes a problem for low supply-voltage operation. This section describes a modified Gilbert cell mixer, which is referred to as a "subharmonic double balanced mixer" in [3]. It has two levels of transistors and an LO at half the frequency of the RF signal.

Figure 6.5 shows a simplified schematic of the double-balanced Gilbert cell mixer. The RF and LO ports are reversed in this circuit. The currents $I_1$ and $I_2$ switch at the rate of LO frequency and are 180° out of phase. These currents are fed to the RF section, and are then mixed with the LO frequencies in the current domain.

First let us illustrate the frequency doubling operation by the principle of collector current combination in the LO section. The phases of the four collector currents, $i_c(Q_1)$, $i_c(Q_2)$, $i_c(Q_3)$, and $i_c(Q_4)$, are given by 0°, 90°, 180°, and 270°, respectively. Current $I_1$, which is the sum of collector currents in $Q_1$ and $Q_2$, switches at twice the LO frequency. A similar doubling operation is valid for current $I_2$, which is the sum of collector currents in $Q_3$ and $Q_4$. The differential LO signal applied to the bases of $Q_3$ and $Q_4$, is 90° out of phase with the LO signal applied to the bases of $Q_1$ and $Q_2$. Hence, current $I_2$ is 180° out of phase with respect to the current $I_1$, due to the frequency doubling effect. This is described in Fig. 6.6.

As $I_1$ and $I_2$ go to the RF section 180° out of phase, double-balanced mixing occurs, and the IF is obtained as a result of mixing the RF signal with the doubled-frequency LO signal. This mixer core uses only two levels of transistors, can be driven using low-voltage power supplies, and can be implemented in a fully CMOS technology. For a CMOS implementation, the bipolar transistors in Fig. 6.5 would

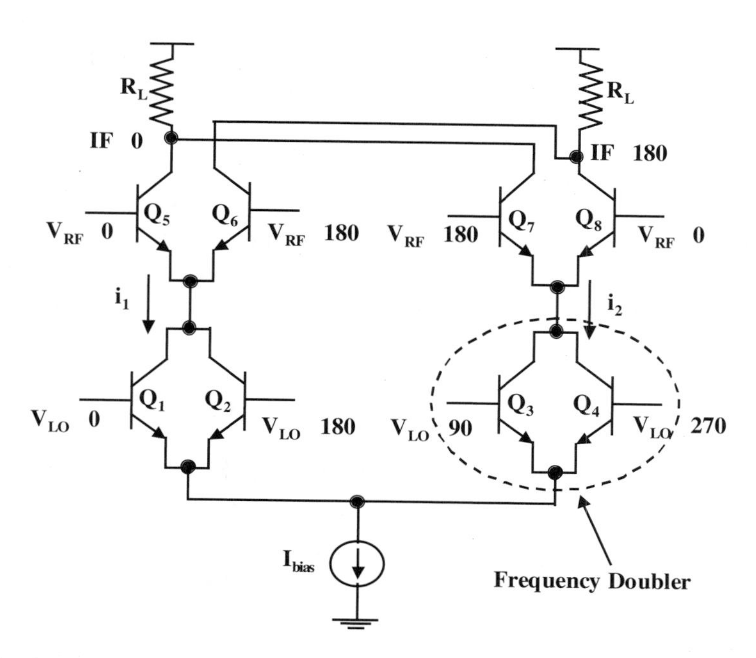


**Fig. 6.5.** Simplified schematic of modified Gilbert cell mixer with parallel stacking.

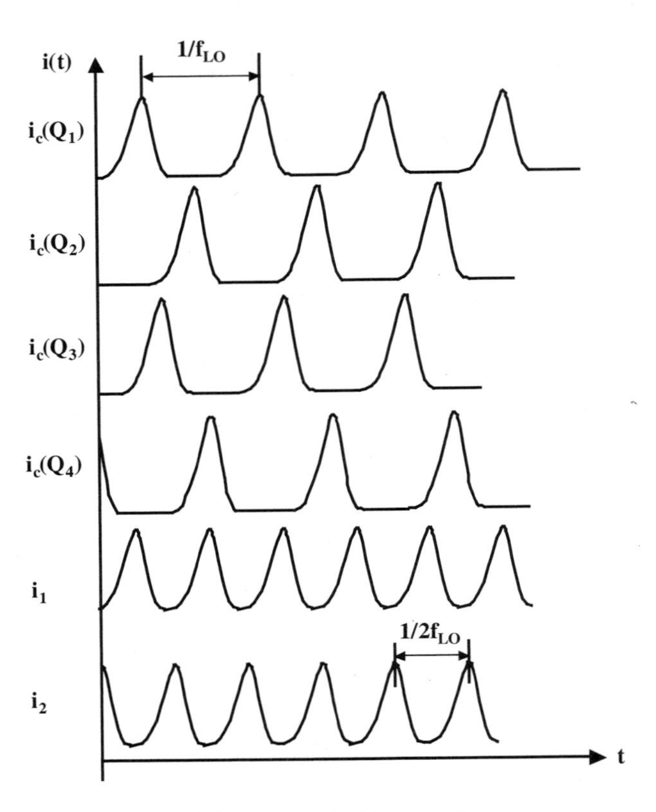


**Fig. 6.6.** Frequency doubler waveforms.

have to be replaced by NMOS transistors. The frequency doubler and the mixing operations remain similar to what has been discussed above for bipolar transistors. With judicious biasing techniques, the power dissipation can be kept low. Thus, this circuit shows potential for the future generation of low-voltage, low-power transceiver subsystem designs.

### 6.2.2. Receiver Architecture

Figure 6.7 shows the receiver architecture under consideration. It uses LO buffers followed by a polyphase filter to generate the appropriate LO phases.

Outputs of the polyphase filter are amplified using the LO buffers. The polyphase filter outputs have phase and amplitude variation with LO input frequency. The two differential LO quadrature signals maintain a 90° phase difference over a wide bandwidth. Inductive degeneration in the RF section can be used to improve linearity of the mixer, as well as input matching. The frequency planning in all the following receivers assumes very low IF (VLIF) or direct conversion architecture.

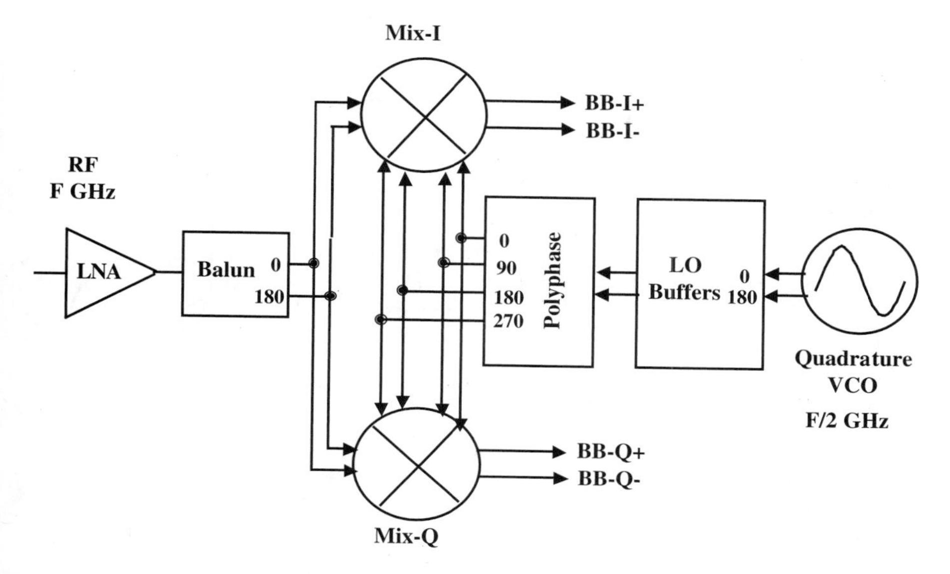


**Fig. 6.7.** Subharmonic receiver architecture using balun, polyphase filter, and LO buffers.

Let us investigate the receiver architecture in more detail. Providing appropriate phase shifting is a very critical decision for this architecture. Assuming a fully balanced architecture in the front-end utilizing both the $I$ and $Q$ channels, there are three places one can provide a 90° phase shifter: (1) input RF signal, (2) LO signal path using polyphase filter, and (3) a 45° phase shifter. Generation of quadrature phases in the RF path can be performed by using a 90° passive phase shifter or a polyphase filter. The differential architecture is shown in Fig. 6.8. One problem associated with this architecture is the loss encountered in the passive phase shifters in the RF path. This loss must be swamped by the LNA gain to achieve a reasonable noise figure of the receiver.

As described above, to generate the appropriate quadrature signals from the mixer, one needs to provide eight LO signals, each differing by 45° in phase from each other (Fig. 6.9). Thus, the doubler core can provide 90° phase signals at twice the LO frequency. These 45° signals can be generated in two ways: (1) using a broadband phase shifter, and (2) directly from the cross-coupled VCOs as described in [5]. The LO buffers should be present to compensate for the loss in the passive phase-shifter networks. However, the amplitude and phase imbalance are concerns in the LO path. The amplitude and phase errors are quite tolerable when generated from the VCO, as opposed to the other methods of quadrature generation. However, this leads to more silicon area due to the presence of inductors in each VCO circuit, as well as increased power consumption in the receiver overall. Doubler-based mixers are quite sensitive with respect to duty cycle errors, and might give rise to a much degraded IIP2.

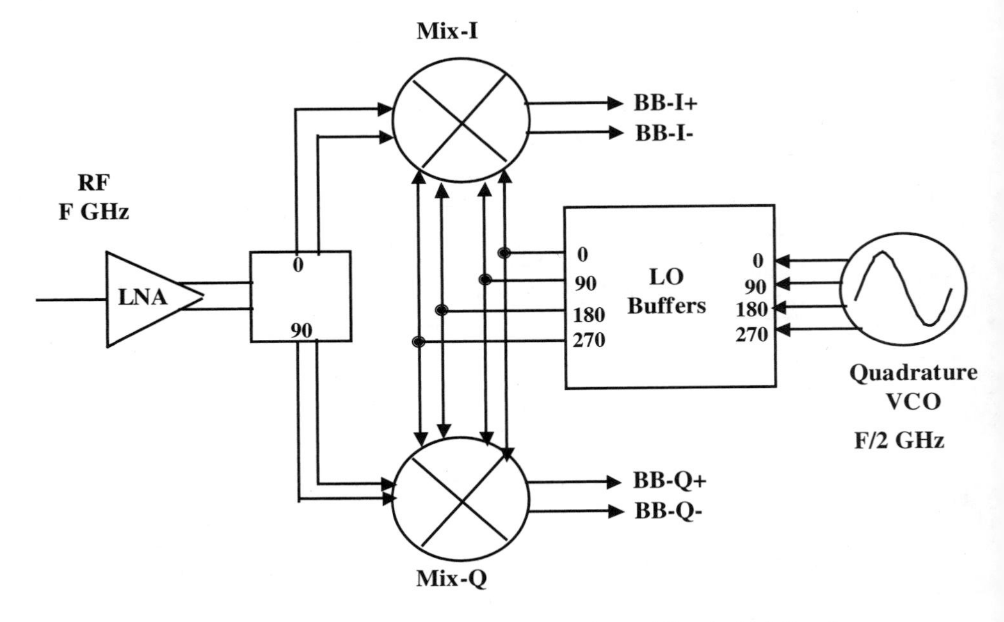


**Fig. 6.8.** Subharmonic architecture using phase shifter in the RF path.

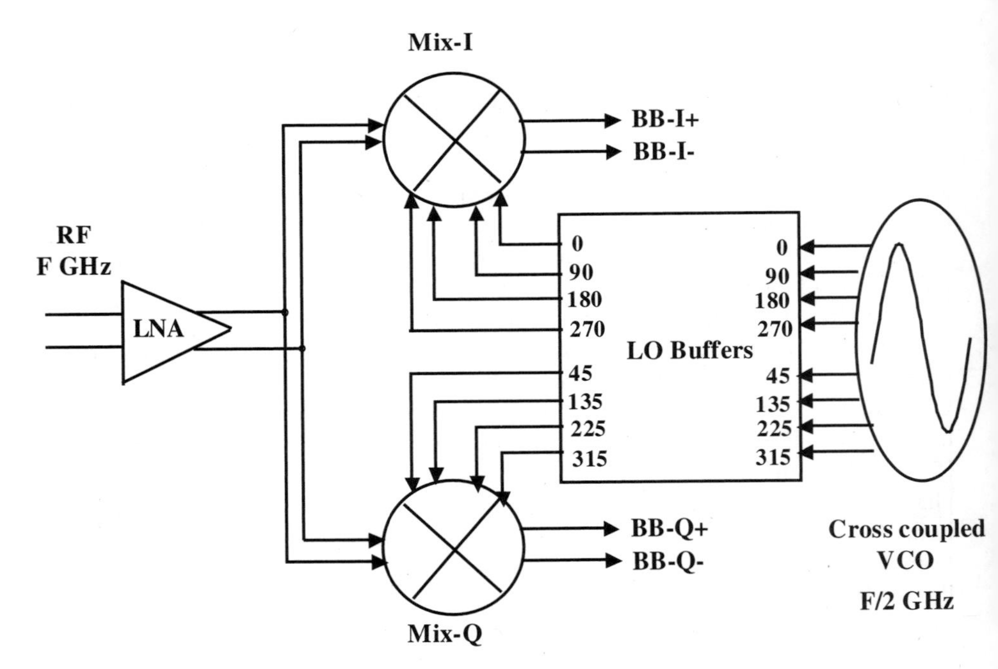


**Fig. 6.9.** Subharmonic architecture using 45° phase-shifted waveform generation from the cross coupled VCO.

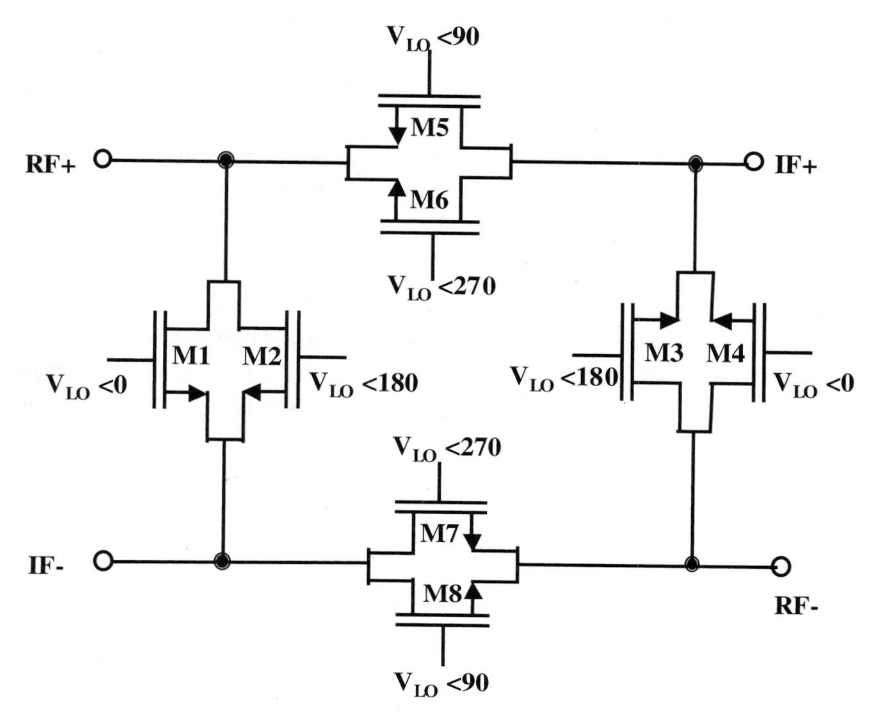


**Fig. 6.10.** Extension of the frequency doubler scheme towards a CMOS passive subharmonic mixer core.

### 6.2.3. Extension to Passive Mixers

The doubler core described above can also be utilized in a passive mixer. The ring-type structure is quite common in CMOS technology due to lower "on" resistance of the MOS transistors. This helps in the formulation of passive mixers in standard digital CMOS technologies. These processes do not provide devices like Schottky barrier diodes and, hence, a doubler-based approach would be useful.

The schematic is shown in Fig. 6.10. The doubler principle works as described before. These types of mixers are quite sensitive with respect to the DC bias voltages applied at the gate and the source terminals.

## 6.3. EXTENSION TO HIGHER-ORDER LO SUBHARMONICS

The previous sections have presented the feasibility of active subharmonic mixing techniques. All of the above techniques use the frequency-doubler-based approach to generate the frequency doubler waveform for the LO signal, and then mix it with the incoming RF frequency. As the frequency of operation increases, it might be quite interesting to investigate techniques by which one could perform mixing operations with higher harmonics of LO tone (this can be referred to as 3×, 4×, etc. mixing, depending on the harmonic order).

Let us consider the serial transistor stack as described in Section 6.1.1. As denoted in Eqn. 6.1, generation of an *N*th harmonic of the LO signal would require *N* transistors in series in the LO switching core of the mixer. This will be clear to the reader by considering Eqn. 6.1 and Fig. 6.1. One interesting feature of this technique is that one can utilize even or odd multiples of the LO tone for subharmonic mixing purposes. This provides additional flexibility in the receiver architecture with respect to LO frequency. One potential problem in this approach is the increase in the number of transistors between the supply rails. This poses a serious problem in terms of technology scaling, especially in the low-voltage application regime.

The above scenario can be improved somewhat by the second approach, which reduces the amount of stacking by stacking transistors in parallel. Figure 6.11

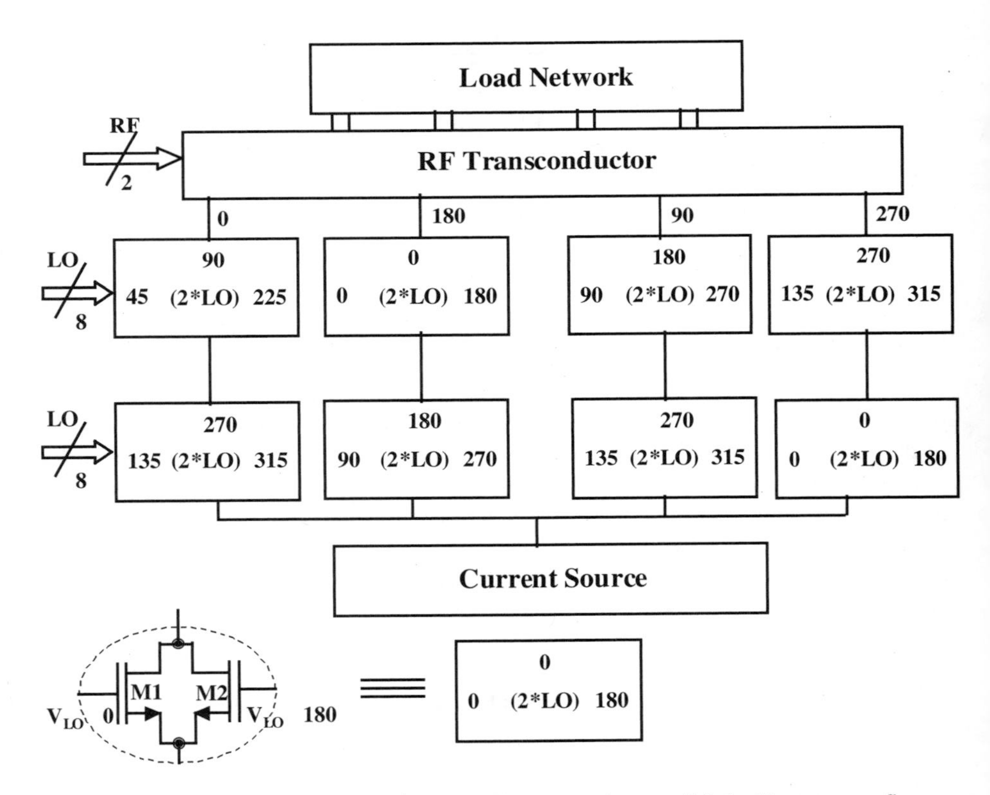


**Fig. 6.11.** Example of a 4× active mixer architecture using parallel doubler core configuration.

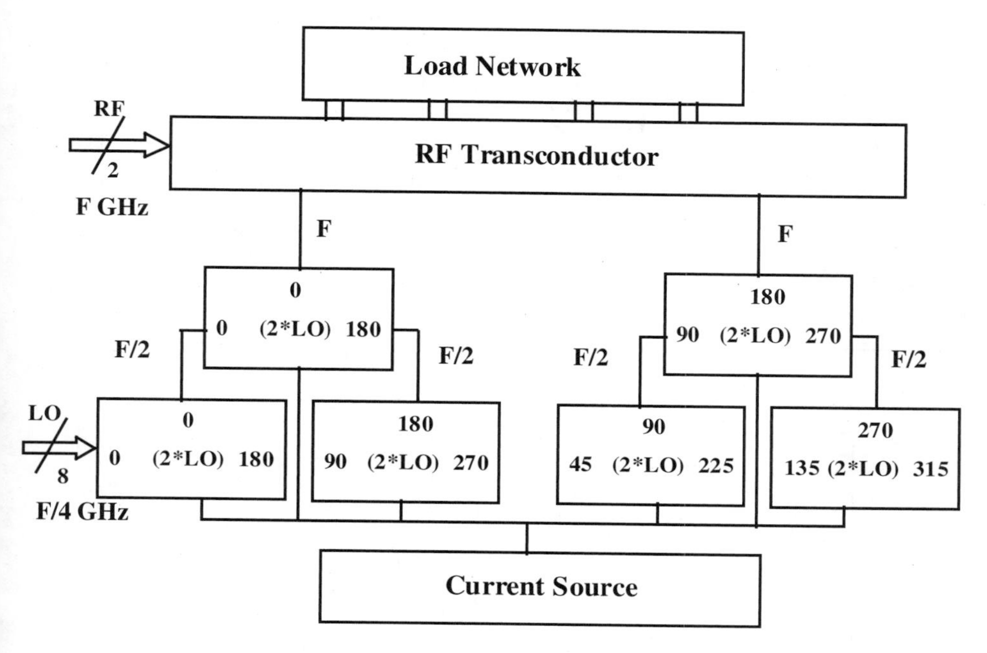


**Fig. 6.12.** Illustration of a 4× active mixer architecture using parallel doubler core configuration, with constant amount of stacking between the supply rails. Frequency planning is for low-IF or direct conversion receivers.

shows a 4× mixing scheme based on the parallel stack doubler core. An $N$th harmonic of the LO signal would require $N/2$ transistors in series in the LO switching core of the mixer. An extension of this approach is shown in Fig. 6.12, where there are three transistors in between the supply rails. However, this requires additional power consumption in the individual doubler cores.

## 6.4. MULTIPLE PHASE SIGNAL GENERATION FROM OSCILLATORS

As described in the previous sections, the subharmonic mixing operation requires multiple phases to be generated on-chip. One possible solution towards multiphase signal generation is presented in [5]; it utilizes a ring-based architecture of cascaded LC oscillators. The configuration is shown in Fig. 6.13. An LC oscillator can be visualized as a very high $Q$ band-pass filter, which filters the side-band noise around the center frequency in a symmetrical fashion. Similar to the operation of the cascaded biquadratic structure of a band-pass filter, the noise filtering effect becomes higher with the increase in the number of identically designed band-pass filters. With the presence of $N$ cascaded identical oscillators, the carrier signal at the resonance frequency is increased by a factor of $N$ and, hence, the signal power increases by a factor of $N^2$. From the noise perspective, with $N$ identical oscillators, the output

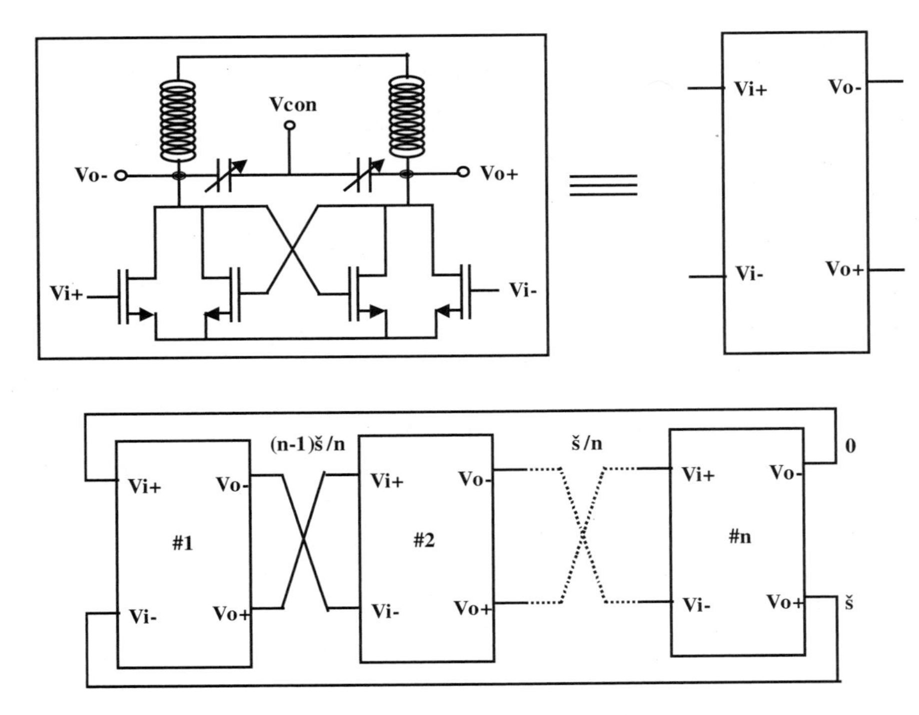


**Fig. 6.13.** Illustration of ring oscillator topology with cross-coupled cores.

noise is reduced by a factor of $N^2$. In the overall effect, the phase noise decreases by $10 \log N^3$ dBc/Hz.

The architecture behaves as a series of cascaded unit oscillators. The ring architecture utilizing $N$ stages has a direct connection and $N - 1$ cross connections. The cross coupling forces the following oscillator into phase, whereas the direct coupling forces it into antiphase. The unit oscillators connected by cross coupling have $\pi/N$ phase difference, and the direct coupled oscillators have $(N - 1)\pi/N$ phase difference.

One attractive feature of this signal-generation scheme is that it does not provide any restriction on the type and topology of unit oscillators to be utilized. However, there are limitations in terms of area in silicon-based technologies, and power consumption in these cascaded stages. Also, the LC ring oscillator is more immune to device mismatch as the number of oscillator stages increases.

## 6.5. FUTURE DIRECTION AND CONCLUSION

In this chapter, we have presented active subharmonic mixers in detail; first, their principles of operation, and then their implementation in subharmonic receiver architectures. Subharmonic receivers are quite attractive for future-generation wireless communication, in which the incoming RF signal will be at a very high frequency and should be down-converted by a relatively low LO signal, to be generated on-chip. As the semiconductor technologies advance, and both the frequency of operation and the data rates increase, the subharmonic techniques become more and more relevant. For 60 GHz receiver designs, subharmonic architectures will play a very important role. Most of the active subharmonic mixing techniques described in this chapter are based on frequency doubler techniques. It would be interesting to see whether an innovative circuit technique can exploit the exponential I–V of devices to get an active subharmonic mixing. This might be a mix between the usual antiparallel diode pair technique, and the standard Gilbert cell topology, and would be capable of handling higher subharmonic mixing using the exponential device characteristics.

## REFERENCES


1. L. Sheng, J. C. Jensen and L. E. Larson, "A wide bandwidth Si/SiGe HBT direct conversion sub-harmonic mixer/downconverter," *IEEE Journal of Solid-State Circuits, 35,* 9, 1329–1337, September 2000.
2. R. Svitek, D. Johnson, and S. Raman, "An active sub-harmonic direct-conversion receiver front-end design for 5–6 GHz band applications," *IEEE Microwave Theory and Technique Symposium Digest, 1,* 505–508, 2002.
3. K. Nimmagadda and G. Rebeiz, "A 1.9GHz double-balanced subharmonic mixer for direct conversion receivers," in *IEEE RFIC Symposium,* pp. 253–256, 2001.
4. B. Gilbert, "A precise four-quadrant multiplier with subnanosecond response," *IEEE Journal of Solid-State Circuits, SC-9,* 365–373, December 1968.
5. J. J. Kim and B. Kim, "A low-phase-noise CMOS LC oscillator with a ring structure," in *IEEE Solid State Circuits Symposium,* pp. 430–431, 2000.

# 7

# DESIGN AND INTEGRATION OF PASSIVES COMPONENTS

## INTRODUCTION

Integration of passive components has been a challenge in the design of RF front-ends for a long time. Since the invention of transistors, digital circuits have steadily continued the trend of integration and followed Moore's law to a great extent as their dimensions have shrunk significantly in the deep-submicron regime. As we have pointed out earlier in this book, the back-end of any transceiver subsystem is a digital processor, which provides the advantage of integration with process scaling. However, it is questionable whether Moore's law is equally applicable to the transceiver front-ends, where integration requires mixed-signal and analog functions. Almost all commercially available wireless solutions today use numerous discrete components in the front-end, which tend to increase size and cost. In this chapter, we introduce the reader to the advances in the areas of passive components. Traditionally, the development of off-chip passive components preceded that of on-chip passives. Off-chip passives tend to exhibit superior performance in terms of their quality factor, which is necessary for front-end circuits. Passive components have a direct impact on the power-added efficiency of power amplifiers and noise and blocker performance in the receiver front-end. There have been numerous reports on development of passive components in ceramic substrates, one such example being low-temperature cofired ceramic (LTCC) material [1–2]. These types of ceramic materials also provide multiple dielectric layers, which helps reduce the lateral dimension of the passive structures, leading to compact size.

With the continued trends in development and integration of semiconductor technologies, use of on-chip passive components for RF front-ends has become more and more desirable. However, at the same time, the shift from GaAs-based

*Modern Receiver Front-Ends.* By J. Laskar, B. Matinpour, and S. Chakraborty
ISBN 0-471-22591-6 

process technologies, which provide semi-insulating substrates, to silicon substrates is creating more difficulties. A major challenge in implementing passives on silicon is the low resistivity of silicon substrates, which has a significant impact on the quality factor of the passives. Poor performance of the passives is one of the major drawbacks in realizing a fully monolithic front-end using silicon-based process technologies. In this chapter, we focus on three different approaches for integration of passives in semiconductor or module technologies:

1. Development of passives in ceramic substrates
2. Development of passives in existing semiconductor technologies
3. Use of additional processing steps to enhance the on-chip passive performance, with an emphasis on fully monolithic implementations.

In this regard, we first introduce the reader to package integration approaches. This is followed by a discussion of on-chip systems, with a detailed illustration of passive components in silicon-based substrates.

## 7.1. SYSTEM ON PACKAGE (SoP)

Depending on the specific wireless standard, several fundamental performance criteria become important: (a) higher operating frequency, (b) wide bandwidth, (c) large dynamic range, and (d) high linearity.

Development of a fully monolithic RF front-end is quite attractive. The fundamental issues in implementing the entire front-end in a single semiconductor technology include: (a) degradation of system SNR caused by signal coupling between the building blocks, (b) lower yield as compared to digital circuits due to process variations, (c) lower quality of passive components, (d) difficulties involved in testing and characterization, (e) fundamental device limitations, an example of which could be device breakdown voltage for a power amplifier design.

The "multichip module" (MCM) approach allows designers to use different semiconductor technologies along with high-quality passive components fabricated on a different substrate, thus achieving superior performance in the front-end. The form factor of such modules could be minimized by using a proper selection of system architecture. There can be three major categories of multichip modules, depending on the materials or processing method of the multilayer stack. These categories are represented as: (a) MCM-L for processes utilizing organic laminate layers, (b) MCM-C for processes utilizing ceramic material (generally co-fired), and (c) MCM-D for processes utilizing thin-film layer deposition. A detailed description of each of these different packaging techniques is presented in [1]. In the system-on-package (SoP) approach, coupling between components through the common substrate is absent. MCM approaches facilitate the use of separate supply voltages for different building blocks and the possibility of integrating MEMS switches and RF filters, along with mixed-signal analog and digital systems, leading to a variety of applications. It should also be noted that solutions that utilize passives integrated in these MCM technologies provide distinct advantages over the surface-mounted discrete components as follows: (a) improved package efficiency; (b) better electrical and high-frequency performance due to reduced parasitics; (c) lower cost, reduced profile, and weight by elimination of a separate package; (d) reduced cost due to elimination of the board assembly process.

MCM-L is usually a low-cost technology option compared to the others. MCM-C is usually between low and medium cost, and MCM-D is usually higher cost. MCM-L evolved from the conventional printed-wire board (PWB) technology, tailored to meet the specific requirements. Epoxy glass is popular as an organic laminate for PWB solution providers. The conductor width, spacing, and via dimensions are on the order of 100–150 αm, similar to regular PWBs. After the assembly process, all the layers are stacked and bonded together by heat and pressure.

MCM-C broadly covers two different technology groups that vary based on the processing temperature. Processing at high temperature is called high-temperature cofired ceramic or HTCC, whereas processing at low temperature is called low temperature cofired ceramic or LTCC. In both of these processes, the ceramic and metal layers are heated at the same time, leading to the term "cofired." In both of these processes, a liquid slurry is formed from ceramic particles and organic binders, which is cast into a solid sheet, commonly referred to as green tape. Afterward, the via holes are drilled and the metal is attached to each layer using screen printing. After screening of patterns, stacked layers are sent to a furnace, where the organic binder decomposes, ceramic densifies, and the structure reacts by shrinking. The firing temperature of LTCC is about 800° and allows the use of noble metals and specific dielectric materials required for integrated passive structures.

MCM-D technology steps are very similar to those of the semiconductor industry, and consist of sequential deposition of conductor (e.g., aluminium or copper) and dielectric layers [e.g., polymide or benzocyclobutene (BCB)] on a substrate base formed of ceramic, silicon, or metal. The spin coating process is used for deposition of dielectric layers to ensure a uniform and controlled thickness. Vias are formed by laser ablation, reactive ion etching, or wet etching of the dielectric material. Thin metal layers are deposited by sputtering and patterned by etching. The curing of the dielectric layers requires lower temperature ranges compared to LTCC. The typical temperature range is 200°C for BCB and 400°C for polymide. Typical geometry parameters for the conductors (width and spacing) in MCM-D technology are on the order of 100 μm.

Coplanar waveguide (CPW) topology plays a major role in designing MCM-based systems due to several advantages such as: (a) ease of connection of shunt devices and circuits, as the connections can be short, thus reducing parasitics; (b) elimination for the need of large drilled via holes through the carrier substrate for ground connection; (c) flexibility in choice of line width for a given impedance; (d) improved shielding by the surrounding ground plane; (e) less influence of material thickness tolerances on system performance.

Having discussed the various types of package technologies, it is time to turn our focus toward the performance of passives in these processes. MCM-L-based passives have poor performance, and are limited to fairly low frequencies. For MCM-C and MCM-D, the passive performance is dependent on the processing technology and the material usage. Due to the superior loss tangent of MCM-D material, this technology exhibits better electrical characteristics in comparison to MCM-C, showing potential for higher-frequency applications. It should be noted that thermo-mechanical reliability problems caused by incompatibility between different LTCC materials can greatly hamper the quality of fabricated passives.

When applying these technologies to frequencies beyond a few gigahertz, it should be noted that the wavelength of the desired signal becomes similar to the dimensions of the passive components. This phenomenon makes use of distributed elements more attractive than their lumped-element equivalents. Illustrations of distributed RF components such as filter, coupler, and balun are provided in [1]. Additionally, for such high-frequency applications flip-chip packaging technologies help to reduce the interconnection parasitics, improving the potential of MCM-based packaging technologies

In the following sections, we describe the development of passive components in multilayer substrates. These structures are developed for minimization of lateral area for each component. The vertical integration of components is a popular technique that uses the multilayer nature of these process technologies to dramatically reduce the lateral dimensions of front-end products.

### 7.1.1. Multilayer Bandpass Filter

Development of a front-end band-pass filter using two resonator structures in a multilayer ceramic process has been reported [5]. The filter utilizes a stripline lumped-element topology to allow compact implementation. A schematic of such a band-pass filter is shown in the literature [5]. This implementation uses six layers of LTCC tape in a stripline configuration. A fully exploded, three-dimensional view of the filter is shown in Fig. 7.1(a), with the cross-sectional view along the line AA′ illustrated in Fig. 7.1(b). Top and bottom metalizations are placed on layers 6 and 0, serving as the top and bottom ground planes. The shunt inductors the $L_1$ and $L_2$ were realized by the U-shaped strips fabricated on layers 4 and 3, which are located two and three layers underneath the top ground plane, respectively. The other ends of the strips are connected to both grounds through vias. Each layer in the process is 3.6 mil thick and no metalization is present between layers 4 and 6; therefore, the top inductor strip on layer 4 is two layers away from the top ground plane, whereas the bottom strip is three layers away from the bottom ground plane. The required mutual inductive coupling is achieved by overlapping the $L_1$ and $L_2$ strips, which are one layer apart. MIM capacitors with electrodes on layers 4 and 3 laid out beside the inductor strips were utilized to implement underneath $C_I$ and $C_O$, as shown in Fig. 7.1(a). The rectangular plates on layers 4 and 3 in Fig. 7.1(a) act as the top and bottom plates, respectively, of underneath $C_I$ and $C_O$. The vertically interdigitated capacitor (VIC) topology was utilized to implement $C_S$, which is realized by two series capacitors of capacitance $2C_S$. Each of these capacitors is implemented in VIC topology as a parallel combination of two capacitors with a value of $C_S$. This is depicted more clearly in the cross-

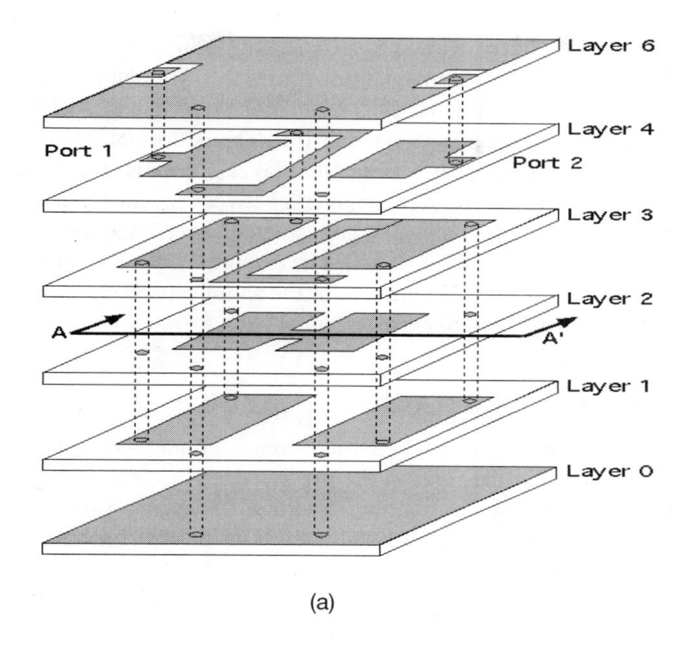


(a)

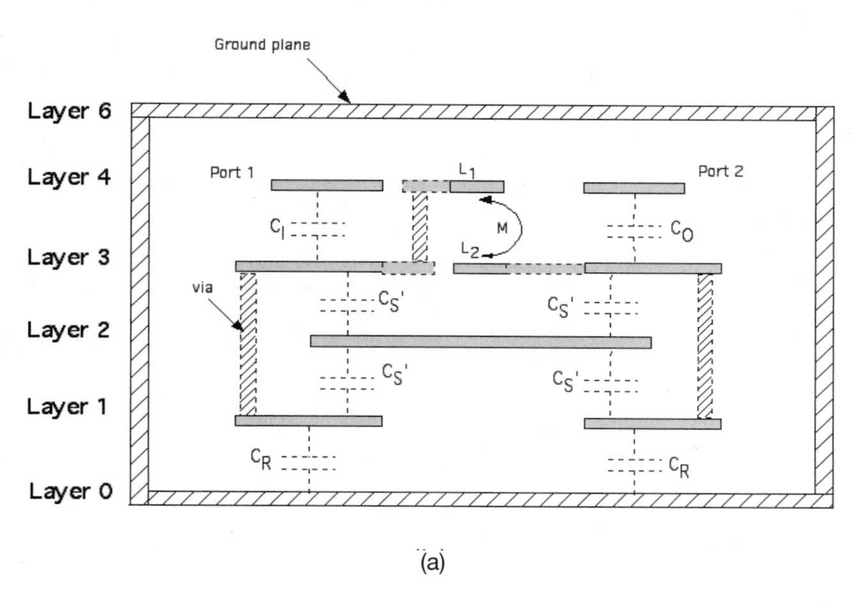


(a)

**Fig. 7.1.** (a) A three-dimensional, layer-by-layer view of a lumped-element, multilayer image-reject filter structure. (b) Cross-sectional view along AA′ indicates how each filter element is deployed.

sectional view in Fig. 7.1(b), showing VICs deployed on layers 3, 2, and 1, right underneath $C_I$ and $C_O$. Such implementation ensures the symmetry of the structure desirable for high-frequency circuits. If $C_R$ were implemented in MIM topology, the entire structure would not have been symmetrical. The bottom plates of $C_I$ and $C_O$ are used as the top plates of the VIC extended to layer 2 through via connections. The "dumbbell"-shaped trace is inserted on layer 2 between layers 3 and 1 as the bottom plates of the VIC. The plates on layer 4 connected to layer 2 and the plates on layer 3 implement $C_S'$, as indicated in Fig. 7.1(b). The extended top plates of the VIC on layer 1 are used as the top MIM electrodes for the shunt capacitor $C_R$ with the bottom ground on layer 0 as the bottom electrode.

### 7.1.2. Multilayer Balun Structure

In this section, we demonstrate a balun developed with a LTCC process similar to the one used to develop the filter in the previous section. The balun has been implemented using a multilayer coupling structure, occupying a total of five dielectric layers, as shown in Fig. 7.2. Two shorted transmission lines (A and B), when placed next to an open-circuited line (C), can couple energy from the open-circuited line, which has a length of half a wavelength at the desired frequency. The standing wave forms a short circuit at the center of the open-circuited line. Here, the current is at its maximum value. At equal distances away from the center and on either side of it, the voltages will be equal in magnitude but of opposite phases. The two short-circuited line sections are coupled from both ends of the open-

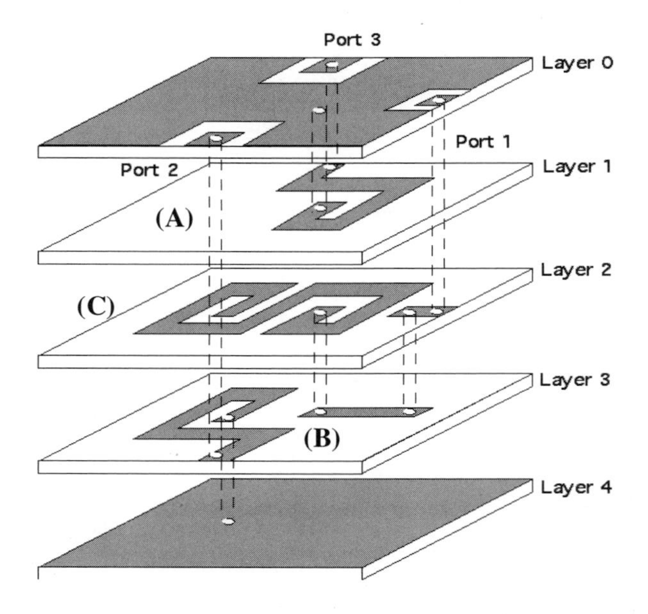


**Fig. 7.2.** Fully exploded three-dimensional view of the multilayer balun structure.

circuited line. Thus, a signal incident at port 1 induces signals at port 2 and port 3 with equal amplitudes but opposite phases. The short-circuited lines will have to be approximately a quarter wavelength long in order to achieve the desired characteristics.

As shown in Fig. 7.2, two ground planes (layers 0 and 4) are connected by vias to ensure that they are at the same potential, and the two coupling sections at the output are on layers 1 and 3 for effective isolation.

### 7.1.3. Module-Integrable Antennas

Different topologies can be adopted for the design of module-integrable antennas. In this section, we briefly describe some of these topologies under the categories of (a) cavity-backed patch antennas (CBPAs), (b) Lifted-slot antennas (LSAs).

***7.1.3.1. Cavity-Backed Patch Antenna (CBPA).*** A CBPA with via feed structure is a potential candidate for implementation in multilayer ceramic substrates such as LTCC. This antenna is targeted for integration with the embedded RF blocks such as filter or duplexer switches. As shown in Fig. 7.3, the cavity structure is formed by surrounding multiple vias connected to the ground plane. The location of via feed is designed to achieve impedance matching with an embedded filter. CBPA needs a smaller ground plane, since the most of energy is confined between the edge of the patch and the via wall. Antenna bandwidth increases with reduction of the gap, with the compromise of decreased radiation efficiency. A thicker substrate increases the bandwidth of the antenna but the thickness should be determined by considering the entire module structure.

***7.1.3.2. Lifted-Slot Antenna (LSA).*** In the substrate, which provides significantly higher loss tangent and thinner dielectric layers, a lifted-slot antenna (LSA) might seem quite suitable. The configuration for an LSA is shown in Fig. 7.4. The inner conductor patch is on the top metal, and the ground plane is located on the conductor at the bottom. When the slot width is much larger than the dielectric thickness, the characteristics of the LSA are similar to those of the conventional slot antenna, except for the feed structure. In the CPW-fed slot antenna, unwanted slot modes are exited at the junction between the slot radiator and the CPW feed. To avoid this problem, via structures can be used as feeds for the patch. Via feed and embedded CPW section can be connected by a mictrostrip line. Location of the via and width of the microstrip line are the design parameters that can be manipulated to obtain impedance matching between the radiator and the feed.

### 7.1.4. Fully Integrated SoP Module

A three-dimensional realization of an integrated module is shown in Fig. 7.5. The separation of the module-integrated antenna has been performed using vertical and horizontal ground planes. With careful system planning, form factors of such modules can be made very small.

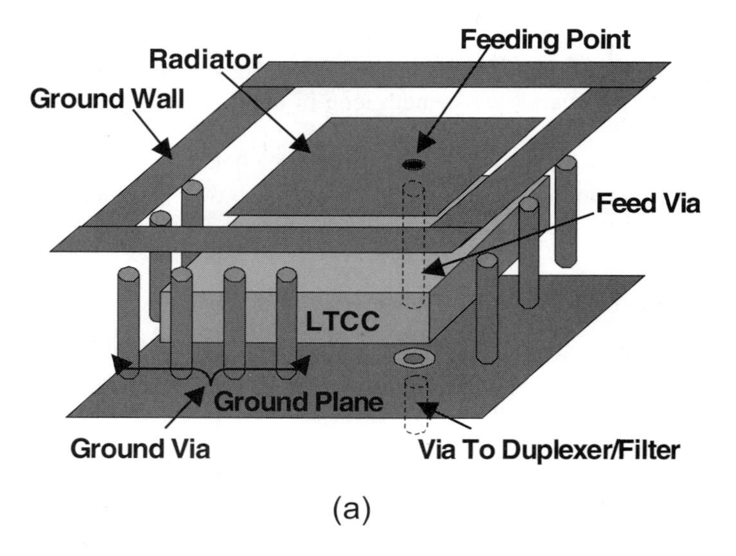


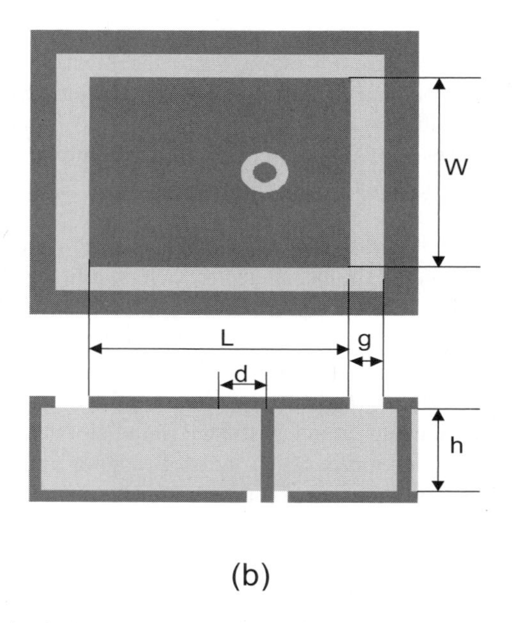


**Fig. 7.3.** (a) Cavity-backed patch antenna is a potential module-integrable antenna. (b) Design geometry parameters for CBPA.

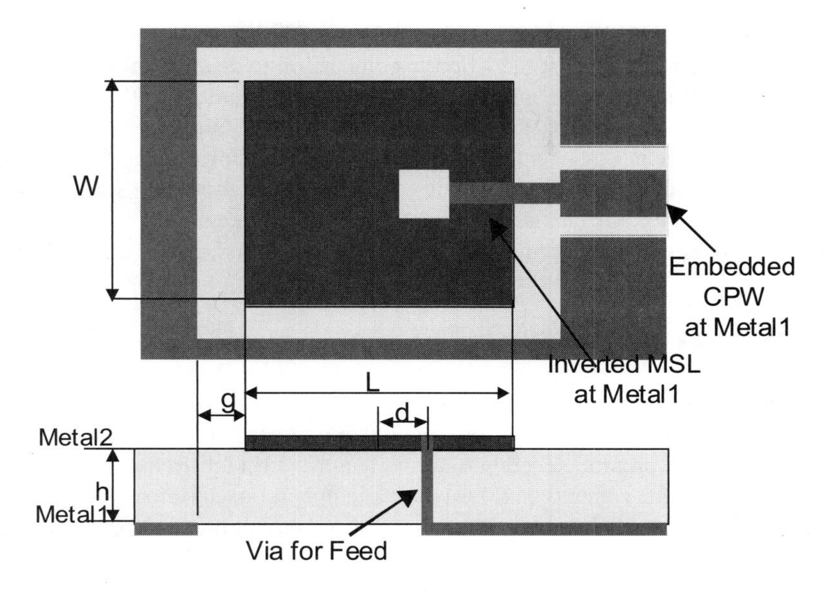


**Fig. 7.4.** Lifted-slot antenna and associated geometry parameters for design.

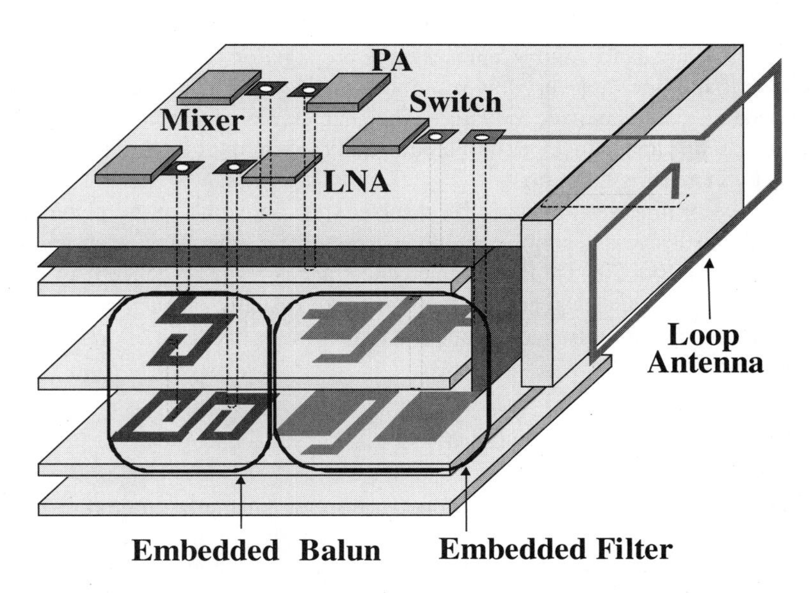


**Fig. 7.5.** Three-dimensional fully integrated system-on-package module.

Although the system-on-package approach has shown great success in performance enhancement and cost reduction in comparison to discrete solutions, this approach still suffers from the inefficiency of integrating multiple technologies. For the ultimate low-cost approach, the integration of actives and passives in a single semiconductor technology remains the most effective approach. In the following sections, we shift the focus of the discussion to the system-on-chip (SoC) approach, and begin to address the challenges associated with the integration of passive components with their active counterparts in a common silicon-based process.

## 7.2. ON-CHIP INDUCTORS

In this section, we describe various aspects of on-chip inductor design. The inductor is one of the most challenging and important components in the RF front-end, as the LC tuned circuits provide several attractive features for high-frequency circuit design, enabling improvements in (a) low-voltage operation, (b) impedance matching between stages, and (c) low dissipation for reduced circuit noise. A detailed discussion of on-chip inductors has been provided in [7].

An inductor can be described as a short section of transmission line whose input impedance is

$$Z_{\text{in}} = Z_0 \cdot \gamma l = (r + j\omega L)l \tag{7.1}$$

This assumes a short-circuit termination for the transmission line. $Z_0$ represents the characteristic impedance, and $\gamma$ represents the propagation constant for a length $l$ of the transmission line. For very low resistivity metals used as interconnects, the input impedance could be made inductive. The characteristic impedance, $Z_0$, is a ratio of inductance to capacitance per unit length, and the length of the inductor could be reduced by increasing the ratio.

Transmission line structures utilize a metal strip above a ground plane, and in typical silicon process technologies the characteristic impedance varyies somewhere between 100–200 Ω. Behavior of substrate is a major concern in silicon-based technologies and depends on the resistivity (usually on the order of 1–100 Ω-cm) and the operating frequency under consideration. The limit of on-chip inductance can be set as [7]

$$\omega L_{\max} < 2\pi Z_{0,\max}\left(\frac{1}{\lambda}\right) \tag{7.2}$$

Thus, the smallest value of on-chip inductance in the gigahertz range (0.5–4 GHz) is about 1 nH. Quality factor of inductors is also an important parameter, and the medium resistivity (1 Ω-cm $< \rho <$ 15 Ω-cm) of silicon substrates restrict the $Q$-factor to numbers below 10. Quality factor of the inductors is critical for the phase noise of oscillators and the noise figure of the LNA in the front-end. The quality factor degradation is caused mostly by conductor losses arising from metallization resistance, the conductive silicon substrate, and substrate-parasitic capacitance.

Several approaches have been investigated for realizing high quality factor inductors in silicon-based substrates [7]. Reported approaches include: (a) use of thick metallization to reduce ohmic losses, (b) use of low-resistivity metals (copper) and stacking of metal layers, (c) use of high-resistivity silicon substrates for fabrication ($\rho > 1$ kΩ-cm), (d) use of micromachining techniques to remove silicon substrate, and (e) use of patterned ground shields.

Inductor geometry plays an important role in achieving a good quality factor. Three commonly used geometries include rectangular, octagonal, and circular shapes. The improvement in quality factor is about 10% from a rectangular to a circular design [7], with a greater area consumption and difficulty in lithography masking. Realizing inductors by means of a square geometry would provide an area advantage over an equivalent straight-wire inductor. However, modeling of such inductors at high frequencies requires analysis of fringing-field effects, parasitics, ground-plane effects, and effects of conductive substrates. Figure 7.6 illustrates a spiral inductor and the associated geometry parameters.

### 7.2.1. Inductor Modeling

Circuit models for inductors in lumped-element form are very essential for design and optimization of RF ICs in the front-end. A scaleable inductor model is very useful for circuit-level optimizations. One major consideration in inductor modeling is

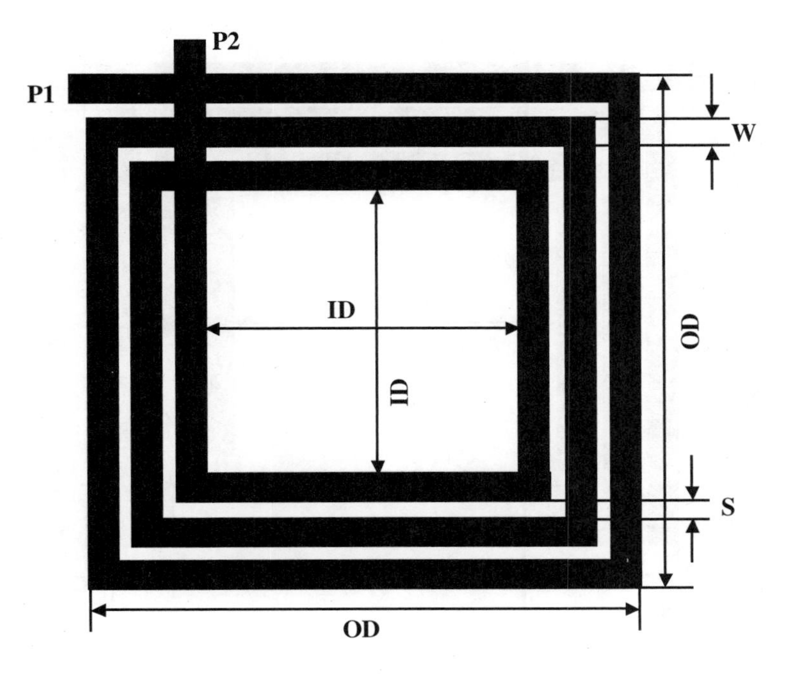


**Fig. 7.6.** Various geometry parameters associated with monolithic inductors.

to account for the ohmic losses caused by three different modes of signal propagation in Si/$SiO_2$-based systems: (1) skin effect, (2) slow wave, and (3) quasi-TEM. Each propagating mode can be represented by the lumped-element equivalent circuits to develop a scaleable inductor model. Shunt-parasitic components of the microstrip line can be represented by a combination of two capacitors and a resistor. Thus, a spiral inductor can be represented as a collection of short microstrip line sections, as shown in Fig. 7.7.

Each of the microstripline segments can be modeled by lumped-element components consisting of series resistance and inductance parameters calculated per unit length. The lumped-element components in a microstripline represent the series resistance, substrate parasitics, and losses. Each of the lumped elements is connected in series to complete the entire model for the spiral inductor. The electric and magnetic couplings between parallel conducting strips should also be captured in the model. Coupling between the orthogonal sections could be ignored to the first order to reduce the complexity of the lumped-element model.

The above approach is commonly described as the "segmented approach" to inductor modeling. This method was proposed by Greenhouse and then extended by Rabjohn [7] for arbitrary configuration of orthogonal microstrip lines. This approach requires complex calculations to quantify the different parameters associated with the inductor model. Consequently, the electrical parameters of this lumped-

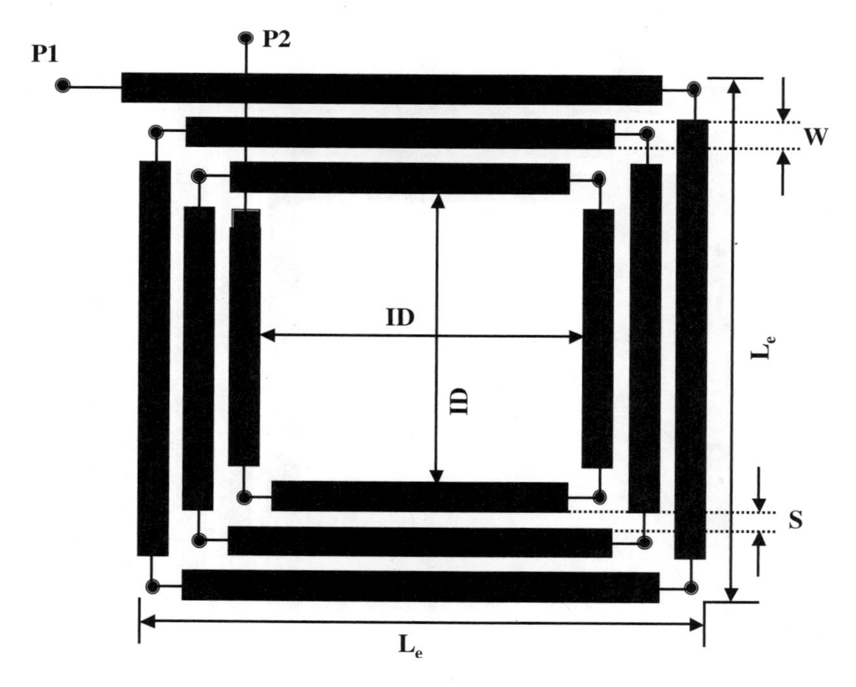


**Fig. 7.7.** Spiral inductor split into several microstripline sections.

element inductor model is calculated using a computer program based on a given inductor geometry and process technology. The self- and mutual inductances of such structures are calculated from parallel line segments in closed-form expressions. The self-inductance, $L$ (in nH) for a straight conductor with rectangular cross section is formulated using Grover's formulation of two current-carrying filaments [8]. The self- and mutual capacitances are computed using two-dimensional numerical techniques for coupled microstrip lines [9]. The shunt components include the substrate resistance and capacitance, represented by $R_{si}$ and $C_{si}$, respectively. The frequency-dependent resistance, $R_{sk}$, is formulated using the closed-form expression provided in [10].

The complete model of a spiral can be formulated after estimating various parameters for each of the microstrip segments. Figure 7.7 illustrates that the physical layout is segmented into groups of coupled lines. First, the parameters of lumped-element $\pi$-equivalent circuits for each individual microstrip lines in each group are extracted. As shown in Fig 7.8, four such equivalent circuits are used to complete a single turn of the spiral inductor. The lumped capacitor between two lines is represented by $C_m$. Dependent current sources are used to account for the mutual magnetic coupling between parallel strips of a group of coupled lines. As the number of turns increases for a given spiral structure, more lumped-element sections are added within each group, and the number of interconnecting elements to model the mutual capacitances are increased.

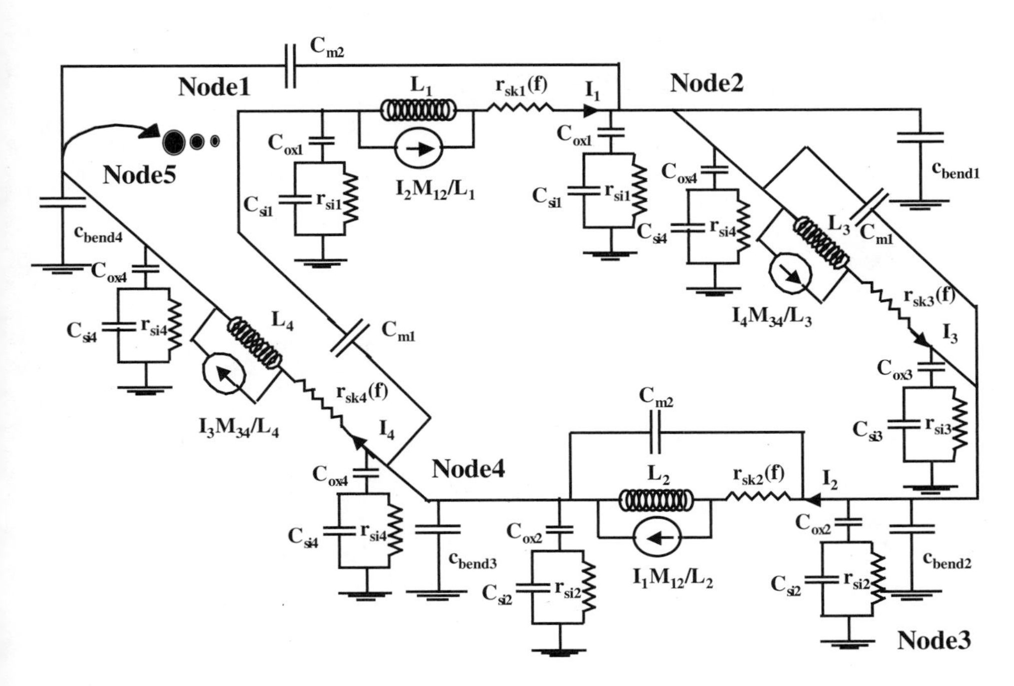


**Fig. 7.8.** Lumped-element representation of a single turn of a spiral inductor.

It should be noted that the explanation provided above for the lumped-element components only accounts for the first-order effects, and this model can be further improved for better accuracy. One additional component that is usually considered in the model is the center contact of the spiral inductor that is implemented as a metal underpass and contributes to the interwinding capacitance. For better accuracy, current-crowding effects that occur at the corners of the spiral structure can be modeled by a parasitic inductance, $L_{\text{bend}}$, and parasitic capacitance, $C_{\text{bend}}$, at each of the corner nodes.

With the introduction of the inductor model and the basic parameters involved, along with their physical interpretations, we can now better understand the circuit model of the spiral inductors. From the circuit design perspective, the model should be able to represent inductors reasonably well over a wide frequency range and be compatible with both time- and frequency-domain simulators. Figure 7.9 illustrates a simplified version of the inductor model in a Pi circuit configuration. Series resistance ($r_s$) and inductance ($L$) of the compact $\pi$ model can be derived by summing the inductance and resistance of individual microstrip lines connected in series. Parasitic capacitances of the individual sections can be combined into a single lumped capacitance by combining all the parasitics associated with each winding (inner and outer) of the spiral structure and representing them at different shunt arms. The shunt parasitics are not symmetric ($C_{\text{ox1}} \neq C_{\text{ox2}}$), due to the asymmetry present in the spiral inductor layout. The $\pi$-equivalent model can also be derived from numerical fit using the characterized data for fabricated monolithic inductor structures. The individual lumped-element components of the $\pi$ model shown in Fig. 7.9 can be optimized using a computer-based optimizer and can be used to model the behavior up to the self-resonating frequency of the inductor [7]. However, a major drawback of the compact inductor model is that the model parameters cannot be easily scaled with changes in the inductor structure, as the components are usually obtained using numerical fit rather than physical modeling.

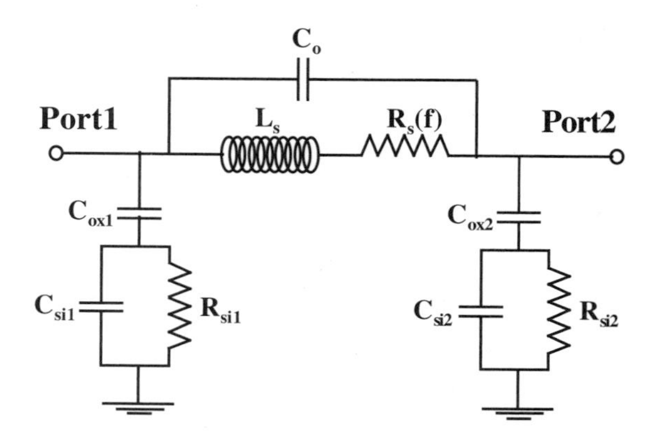


**Fig. 7.9.** Simple reduction of spiral model to a Pi network.

### 7.2.2. Inductor Parameters

Two important performance parameters for monolithic inductors are self-resonating frequency (SRF) and quality factor ($Q$). The quality factor ($Q$) represents the ratio of inductive reactance ($\omega L_s$) to the total series resistance ($r_T$). The $Q$ factor can be measured in a single-port or a two-port configuration. In a single-port measurement, the $Q$ could be estimated by the ratio of the imaginary to the real part of the input port impedance. This is usually accurate for lower frequencies (<500 MHz). Error is introduced by parasitic capacitances as frequency increases, but this can be avoided by estimating the $Q$ value directly from the compact $\pi$ model. When the $\pi$ model parameters are estimated accurately, the $Q$ factor of the inductor can be formulated by grounding the second port, and, from Fig. 7.9, can be expressed as [7]

$$Q \approx \frac{\omega L}{r_s + \dfrac{\left(\dfrac{\omega}{\omega_{ox}}\right)^4 r_{si1}}{1 + (\omega C_{ox1} r_{si1})^2}} \tag{7.3}$$

where, $\omega_{ox}$ is the oxide resonant frequency defined by $L$ and $C_{ox1}$. The total resistance includes the series resistance as well as the added dissipation from the conductive substrate. It is also seen that decreasing $C_{ox}$ with the help of a thicker oxide would increase the oxide resonant frequency, $f_{ox} = (1/2\pi C_{ox1} L)$, as well as substrate corner frequency, $f_{sub} = (1/2\pi C_{ox} r_{si1})$, implying an improvement in the overall $Q$ factor. Use of a higher-resistivity substrate would reduce the $Q$ factor at frequencies much lower than the substrate corner frequency ($f \ll f_{sub}$) and improve the $Q$ factor at frequencies much higher than the substrate corner frequency ($f \gg f_{sub}$). Equation 7.3 is reasonably accurate for a limited range of substrate resistivities, and not accurate in cases of highly conductive substrates ($\rho < {\sim}1$ Ω-cm). The shunt-parasitic effects are not significant at frequencies much lower than the peak inductor $Q$ frequency, leading to an increase in inductive reactance and inductor $Q$ with frequency.

With the increase in frequency, the AC resistance and the dissipation of energy in the semiconducting substrate increase faster compared to the inductive reactance. This causes the $Q$ factor to increase initially, peak, and then decrease as frequency increases. With the increase in surface area, and use of wide conductor metallization, the parasitic capacitance increases, leading to reduced inductor SRF and increased substrate dissipation. Nonuniform current distribution caused by the skin effect also contributes by resulting in higher AC resistance at higher frequencies. Use of wider metal strips can worsen this skin effect, causing a shift in the in the peak inductor $Q$ frequency. A study of inductor performance based on different geometry parameters has been performed in [7]. Study shows that narrow line spacing increases magnetic coupling between windings, resulting in higher inductance and $Q$ factor for a limited area.

Increase in conductor thickness leads to an improvement in the $Q$ factor, as the losses caused by finite metallization resistivity and conductive substrates reduce.

This is often a limitation on achieving higher $Q$ factor in standard digital technologies. In such processes, $Q$ factor could be thought to be proportional to the metal thickness, due to its direct effect on the resistance. The value of inductance is inversely proportional to the thickness of the conductor. The skin effect also predominantly lowers the quality factor from the expected result. Substrate is a major element in inductor performance. Metal layers are usually isolated from the substrate by an oxide layer. A thicker oxide for the insulating layer reduces the parasitic capacitance of the inductor, improving the SRF.

### 7.2.3. Application in Circuits

Having discussed the various aspects of inductor modeling and performance parameters, we now focus on the effects of the inductor on front-end circuits. One of the major concerns of fully monolithic RF front-ends is RF matching, and the quality factor of the inductor plays a major role in this component. One of the significant impacts of a lower quality factor inductor is observed at the input of the LNA, where the inductor loss at the input can degrade the noise figure of the LNA and, consequently, increase the noise performance of the entire receiver chain. Depending on the sensitivity and bit error rate of the system, the noise figure can be a very important criteria. Another implication of lower quality factor of inductors is the poor phase noise performance of the on-chip VCO. This can directly affect the receiver performance by increasing the levels of in-band noise for low-IF and direct conversion receivers, and leads to significant BER degradation. Other parameters that can be impacted by the quality factor of the inductor are the bandwidth and insertion loss.

Apart from the cases discussed above, it is preferable that the circuits utilize a lower quality factor value, as broad bandwidth is preferable in many cases. In the case of circuits providing very high quality factor in their frequency response, a little drift in one of the component values creates a significant degradation in performance. One such example is the LO buffer, which is provided in between the VCO and the mixer for isolation. In the case of bipolar transistors, the input capacitance is relatively small and an inductive load is not necessary. However, in the case of a MOS input stage, as the input capacitance is significantly higher, the use of inductive load becomes quite attractive, as it helps cancel out a part of the input gate capacitance. If the tuning inductor provides a very high quality factor, a minor deviation in the associated circuit elements might lead to a severe mismatch and a lower voltage swing at the LO port of the mixer, leading to poor conversion characteristics. In such cases, a lower quality factor of the inductor is quite desirable.

As described earlier, the on-chip inductor still remains a limiting factor for better performance and lower area consumption of RF circuits. Area used by single inductors poses significant space consumption for RF circuits. In addition, in an integrated front-end, the inductors must be placed at some distance from each other to obtain adequate isolation, leading to even larger area requirement. Adjacent inductors must be spaced far apart to avoid problems such as unwanted coupling between two different circuit nodes.

## 7.3. CAPACITORS

Like inductors, capacitors are one of the basic elements in RF ICs. Different types of capacitors are used in semiconductor technologies: (a) gate capacitors; (b) junction capacitors; (c) metal-to-metal, metal-to-poly, and poly-to-poly capacitors; and (d) thin-insulator capacitors. Among the above capacitor types, the gate capacitors provide highest capacitance density (capacitance per unit area) but they are nonlinear in nature and require a DC bias voltage for operation. Gate capacitors tend to exhibit a relatively medium quality factor and a low breakdown voltage due to the thin gate oxide. Junction capacitors also tend to exhibit similar problems. They are quite nonlinear in nature and need a DC bias voltage for their operation. In addition, junction capacitors are quite sensitive to process variations, and exhibit large temperature coefficients. Metal-to-poly and metal-to-metal capacitors are more linear and provide high quality factors and small temperature variations. However, the density for metal-to-metal capacitors is quite low due to the relatively thick interlayer oxide. With technology scaling, this becomes a problem, as the vertical spacing of the metal layers remains relatively constant. This leads to large area consumption for parallel plate capacitors, a trend opposite to the technology scaling for the active devices. Double poly capacitors and metal–insulator–metal (MIM) capacitors utilize a thin oxide to achieve high capacitance density. The capacitance density achieved by using thin-insulator capacitors is much higher than that of a metal-to-metal capacitor, but lower than the gate capacitor of the same technology [19]. Double poly capacitors and MIM capacitors are highly linear in nature, but require additional masks and processing steps. Hence, they are not commonly available in standard digital CMOS process technologies. MIM capacitors are preferred in narrowband applications, such as filters and resonators, where a high quality factor is needed, whereas the MOS capacitor is a better choice when a large value of capacitance is required. However, the bottom plate parasitic capacitance often tends to be quite detrimental in terms of high-frequency applications.

With increase in performance and reduction in the size requirements of new wireless systems, high-density linear capacitors have become a necessity for modern IC designs. Novel approaches for realizing high-density capacitors are presented in [19]. Figure 7.10 illustrates a lateral flux capacitor. Two terminals of this capacitor use the same metal layer. As technologies scale, lateral fringing becomes more important than vertical fringing effects. The lateral spacing between the metal layers, $s$, shrinks with technology scaling, but the metal layer thickness and the vertical spacing of the metal layers, $t_{ox}$, remain almost constant. This implies an advantage relative to capacitor density in lateral flux capacitors and forms the basis of high-density capacitance. Figure 7.11 shows the simultaneous use of lateral flux and vertical flux to increased capacitor density. As the fringing effect is proportional to the periphery, a large periphery in a specified area tends to increase the capacitance value. One proposed method for realizing high values of periphery in a given area is by using fractals [19]. In this section, we describe the formation of different high-density, on-chip capacitances for front-end development.

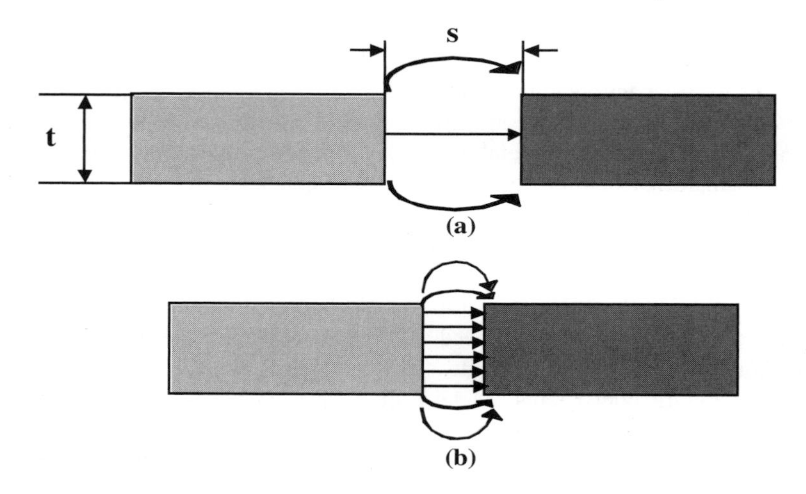


**Fig. 7.10.** (a) Electric field flux capacitor. (b) Effect with technology scaling.

Figure 7.12 illustrates a fractal geometry formation. A fractal can be formed and characterized by starting with an initiator geometry, then replacing each of the edges of the initiator with a curve called a generator. An ideal fractal is defined as a geometrical shape of infinite periphery in a given area. Figure 7.12 illustrates a fractal formation in which the initiator is a square with $M = 4$ sides. The construction of the fractal would continue by recursively replacing each segment of the ini-

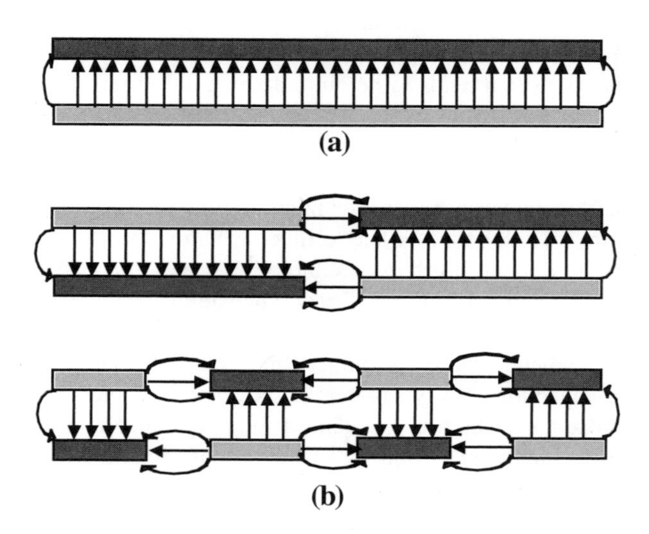


**Fig. 7.11.** Vertical and lateral flux mechanisms. (a) Conventional parallel plate. (b) Cross connection.

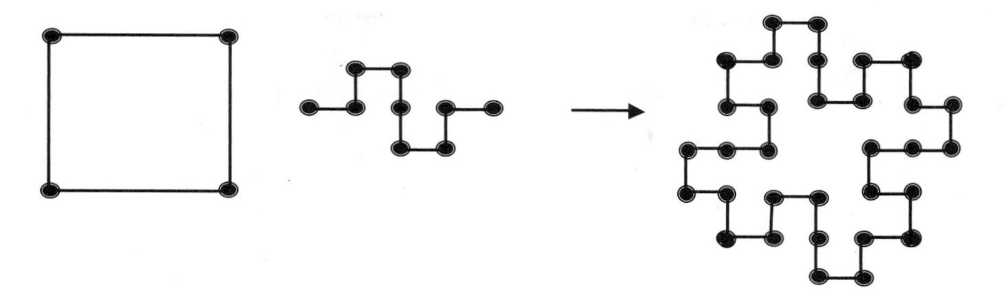

**Fig. 7.12.** Illustration of fractal formation using initiator and generator.

tiator with a curve called a generator comprised of $N = 8$ segments. The size of each generator is $r = \frac{1}{4}$ of the initiator for this example. The fractal dimension, $D$, measures the complexity of a fractal and is formulated as

$$D = \frac{\log(N)}{\log(1/r)} \tag{7.4}$$

where $N$ is the number of segments of the generator and $r$ is the ratio of the generator segment size to the initiator segment size.

Although fractals are quite helpful in lateral flux capacitors, lithographical restrictions are deterrents for implementing these structures in standard semiconductor processes. A realization of a fractal capacitor is illustrated in Fig. 7.13.

One advantage of flux capacitors and fractal capacitors is the reduction of bottom plate capacitance. This is quite useful for high-frequency applications and is illustrated in Fig. 7.14. Due to the higher density of capacitance, the area requirement is less for fractal capacitors. Also, some of the field lines originating from one of

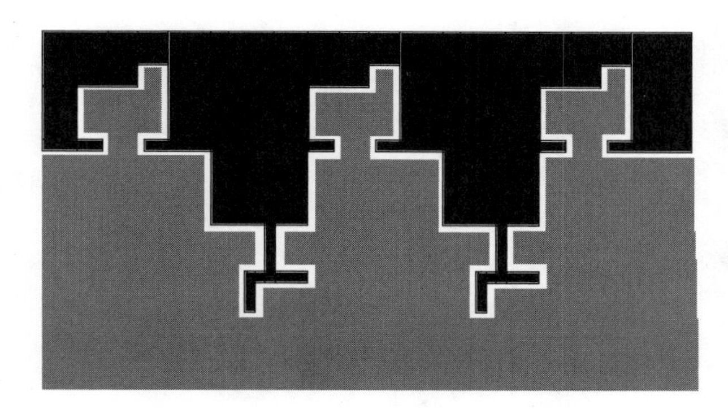

**Fig. 7.13.** Illustration of physical layout of fractal capacitors. Two terminals of these capacitors are represented with different shading.

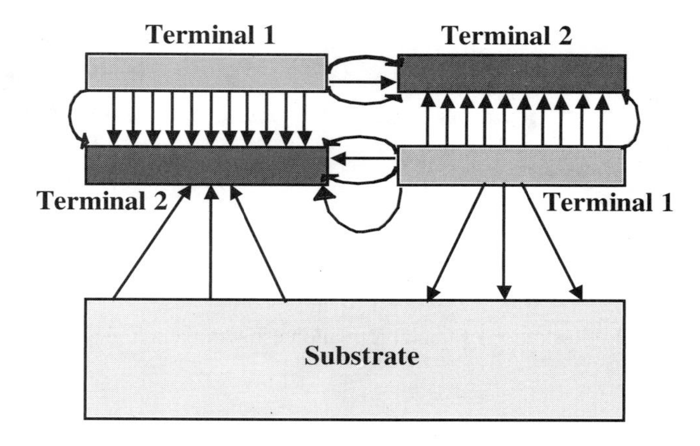


**Fig. 7.14.** Termination of field lines on adjacent plates and reduction of bottom plate capacitance.

the bottom plates terminate on the adjacent plate instead of the substrate, which further reduces the bottom plate capacitance.

Another method for increasing capacitor density is illustrated in Fig. 7.15 in the form of an interdigitated capacitor. This structure, however, gives rise to some parasitic inductance, as the direction of current flow is the same in all parallel stubs. In fractal geometries, the direction of current is randomized, thus leading to a much lower series inductance.

Another way of achieving high-density capacitance is to use the woven structure shown in Fig. 7.16. Different metal layers should be used for connecting to vertical and horizontal lines. Parasitic inductances are lower compared to the interdigitated structure. Also, series resistance contributed by the vias is smaller compared to the case of the interdigitated structure.

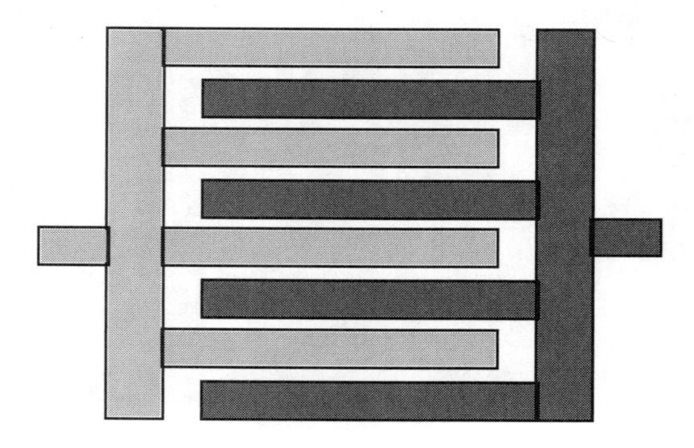

**Fig. 7.15.** Interdigitated capacitor structure to increase capacitance density.

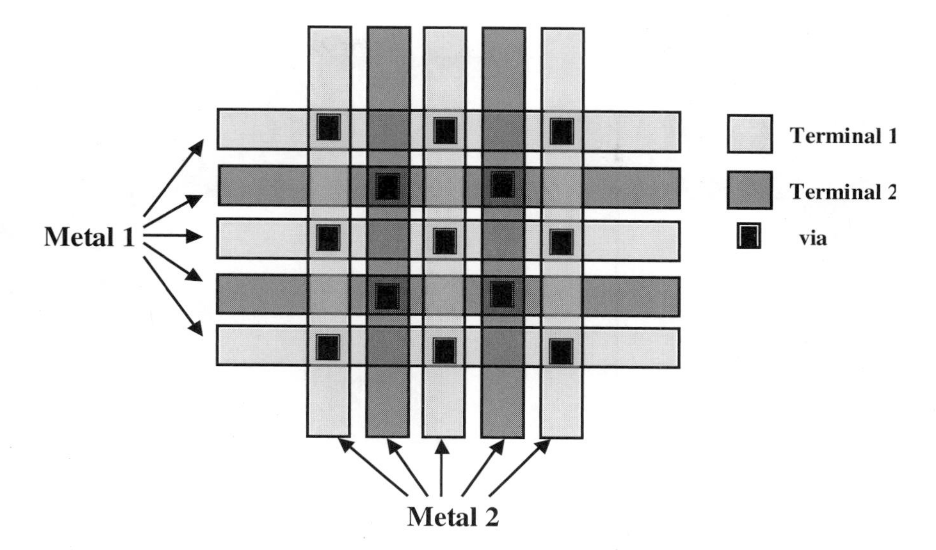


**Fig. 7.16.** High-density capacitance implementation using woven structure.

## 7.4. DIFFERENTIALLY DRIVEN INDUCTORS

Having discussed the inductors and their various parameters associated with circuit design, we now focus on differentially driven inductor structures. Differential inductor structures are quite attractive due to their higher quality factor ($Q$), layout symmetry, and area compaction, which is essential for a miniaturized solution for commercial wireless applications. They are also superior compared to their single-ended counterparts due to their enhanced noise performance.

A detailed description of differentially driven inductor is provided in [11]. Figure 7.17 shows a differential inductor structure consisting of two single-ended inductor structures. It is interesting to note that although the structure is differential in nature, the excitation is single-ended when applied to the structure. AC currents at the input and output ports flow in opposite directions and, hence, some physical separation is needed between the two single-ended structures to limit the mutual magnetic coupling between the two inductors.

Figure 7.18 illustrates a fully symmetric differential inductor structure with differential excitation. This structure can be developed from our earlier understanding of microstripline segmentation of a spiral structure. This is accomplished by joining groups of coupled microstrips from one side of the axis of symmetry to the other using a number of cross-over and cross-under connections. Initially proposed by Rabjohn [11], this structure uses both electrical and geometrical symmetry, and uses the same area as two separate single-ended inductors. The common node of the structure can be utilized as a common mode biasing point for interfacing with circuits. From our earlier discussions, migration to a fully differential structure provides a

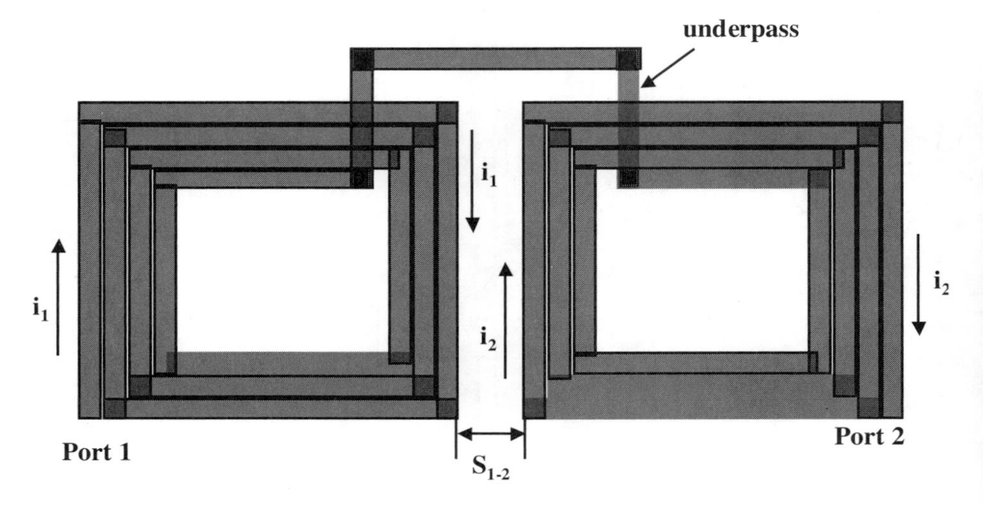


**Fig. 7.17.** Two single-ended inductors connected together in a differential fashion.

significant advantage in terms of space reduction as compared to two single-ended asymmetric inductor structures. The two differential ports are at the same side of the inductor structure, thus aiding integration and layout compaction while designing with active circuits.

The differential inductor structure achieves higher $Q$ than its single-ended counterpart. Two different lumped-element network models are shown in Fig. 7.19 to il-

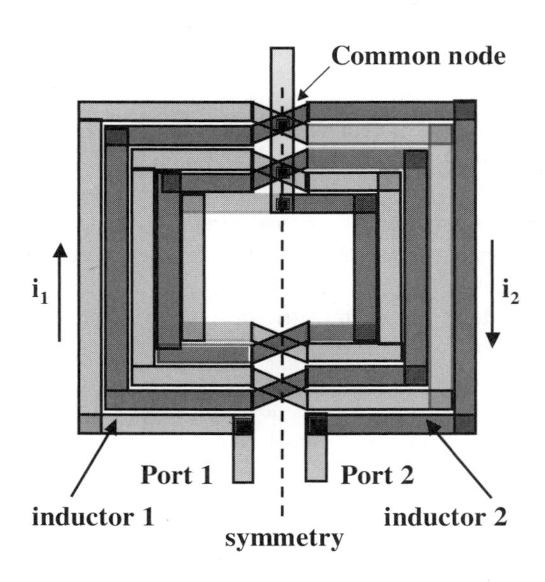


**Fig. 7.18.** A symmetric differential inductor with two ports.

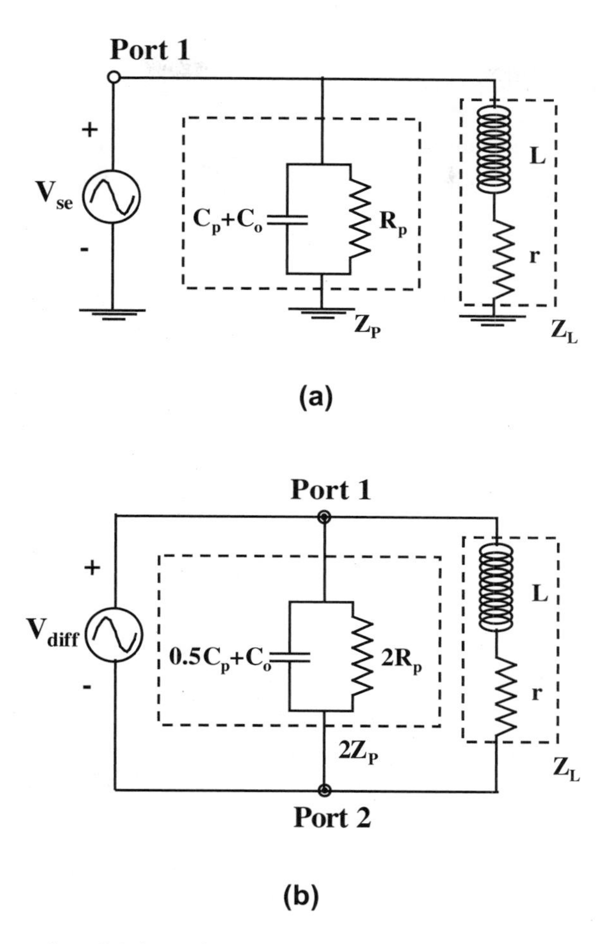


**Fig. 7.19.** Illustration of $Q$ factor for (a) single-ended excitation, (b) differential excitation.

lustrate the improvements in the case of differential excitation. $Z_L$ denotes the impedance corresponding to the inductance and series resistance ($L$ and $r$). $Z_P$ denotes an equivalent shunt-parasitic $R$–$C$ network that provides the same impedance as substrate-parasitic elements $C_{ox}$, $C_{si}$, and $R_{si}$ at a given frequency. For a single-ended excitation, the input signal is applied across the inductor as a one-port structure, and the input impedance $Z_{se}$ is denoted by the parallel combination of the branch impedances ($Z_L$ and $Z_P$). For differential excitation, signal is applied between the two ports, and the differential input impedance is represented by a parallel combination of $Z_L$ and $2Z_P$. A higher equivalent shunt impedance is obtained in case of differential excitation, and $Z_d$ approaches to $Z_L$ over a wider range of frequency than $Z_{se}$. At lower frequencies, the input impedance in either shunt or the differential connections is almost the same, but at higher frequencies, the substrate parasitics ($C_P$ and $R_P$) contribute to the impedance expressions. In the case of differential excitation, the impedance presented by these parasitics is higher compared to the single-ended counterpart at a given frequency. Thus, the real part of the input impedance is reduced and the reactive part is increased, leading to an improvement in $Q$ value. The SRF also increases due to the effective reduction of parasitic capacitance from $C_P + C_o$ to $C_P/2 + C_o$. The ratio of differential $Q$ to single-ended $Q$ is given by [11]

$$\frac{Q_d}{Q_{se}} = \frac{2R_P||R_L}{R_P||R_L}, \qquad R_L = r(1 + Q^2), \qquad Q_L = \omega L/r \tag{7.5}$$

At low frequency ranges, $R_P \gg R_L$, leading to $Q_d \approx Q_L$. $Q_L$ dominates in both the cases, and the $Q$ factor increases with an increase in frequency. At higher frequencies, $R_L$ increases while $R_P$ decreases, leading to an effective increase in $Q$ factor in case of differentially excited structures. From Eqn. 7.5, it is seen that the $Q$ factor improvement in the case of differentially driven structures is about twice that of the single-ended structures. This improvement can also be observed in the case of other passive structures such as couplers, hybrids, and transformers.

## 7.5. TRANSFORMERS

The transformer is another one of the major components in the RF front-end, and its implementation dates back to the early years of wireless telegraphy. A fully monolithic implementation facilitates integration with other building blocks of the front-end. In this section, we describe the various design aspects associated with transformers, along with their modeling and circuit applications. A detailed study of silicon-based monolithic transformers has been presented in [12]. Transformers couple AC current from one winding to another without significant loss in power. The energy transfer and impedance transformation is defined by the mutual inductance between two or more conductor sections. The two different coils of transformers are isolated when considering the DC currents and voltages. Transformers help transform the impedance between primary and secondary terminals, depending on the turn ratio of the two windings. Figure 7.20 shows the geometry of a spiral transformer, and the same analogy as used in the segmented approach can be used to formulate and analyze transformers.

### 7.5.1. Electrical Parameters

Figure 7.21 shows the electrical equivalent of the transformer circuit. AC current flowing in the primary, $i_P$, creates a magnetic flux in the primary coil. $P$ creates a current in the secondary winding, which flows out of the terminal producing a positive voltage, $v_S$, across a load connected between the two terminals $S$ and $\overline{S}$. The electrical parameters of transformer are denoted by the number of turns, $n$, and the

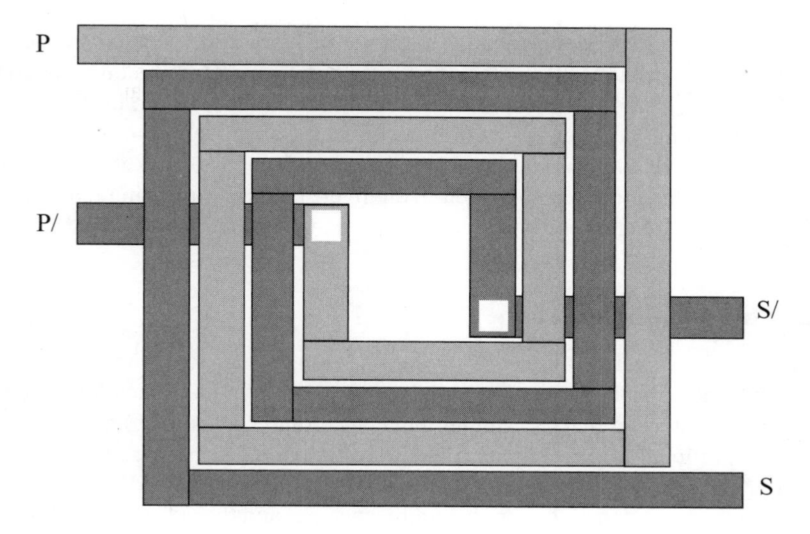


**Fig. 7.20.** Spiral monolithic transformer.

magnetic coupling coefficient $k_m$. Current and voltage at each of the windings in a transformer can be related by

$$n = \frac{v_S}{v_P} = \frac{i_P}{i_S} = \sqrt{\frac{L_S}{L_P}} \tag{7.6}$$

where $L_P$ and $L_S$ denote the self-inductances of the primary and secondary windings, and $v$ and $i$ denote the voltage and currents at the individual terminals, primary and secondary, as shown in Fig. 7.21. The strength of magnetic coupling between the windings is indicated by the $k$ factor, which can be expressed as

$$k_m = \frac{M}{\sqrt{L_P \cdot L_S}} \tag{7.7}$$

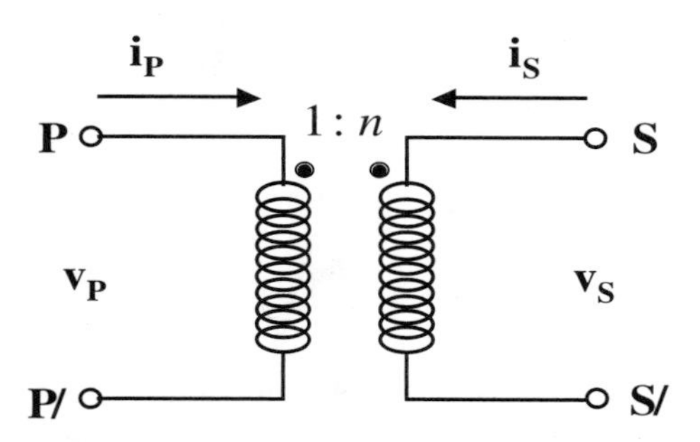


**Fig. 7.21.** Equivalent circuit of a transformer.

where $M$ is the mutual inductance between the primary and secondary windings. The $k$ factor for a practical transformer ranges between zero and unity. Magnetic properties of the materials in IC processes are similar to those of air, leading to $M < \sqrt{L_P \cdot L_S}$. Coupling coefficients as high as 0.9 can be realized on-chip. The input and output phase relationship of the induced voltage depends on the choice of reference terminal. The output can be taken in an inverting or noninverting manner, and has a significant effect on the bandwidth of the transformer.

### 7.5.2. Physical Construction

Figure 7.22 shows various kinds of monolithic transformer structures. Monolithic transformers are constructed by interwinding conductors in the same plane or overlaying them as stacked metals. Mutual inductance (and capacitance) of such structures is proportional to the peripheral length of individual windings. The magnetic coupling coefficient, $k_m$, is determined from the mutual and self-inductances of the windings, which are dependent on the width, spacing, and the substrate thickness of the semiconductor process under consideration. Similar discussions of inductor structure optimization could also be used in the case of transformer optimization. In this section, we describe some of the physical geometries and parameters related to transformer performance.

Figure 7.22(a) shows a Shibata coupler structure [13], in which two parallel conductors are interwound in the same plane for improving edge coupling for the magnetic field. It should be noted that as the layout is asymmetric in nature, with equal numbers of turns in the two windings, the entire lengths of primary and secondary windings are different.

Figure 7.22(b) shows a Frlan-type transformer structure [14]. The transformer terminals in this layout are on opposite sides, facilitating the use of such structures in conjunction with circuit elements. Figure 7.22(c) shows a Finlay transformer [15], which utilizes multiple stacked metal conductors to realize edge and broadside magnetic coupling and reduce area. In a Finlay transformers, lower dielectric thickness improves the performance by enhancing magnetic flux linkage. The lower winding shields the upper winding from the substrate, giving rise to different parasitic capacitances for each winding. Frequency response of such structures is affected by the parallel plate capacitance formed by overlap of the metal layers. Fig. 7.22(d) shows a single-turn coupling transformer; the common periphery between the two windings is limited to only a single turn. These transformers are used in designs in which a low ratio of mutual inductance to self-inductance is desired.

### 7.5.3. Electrical Models

In this section, we describe transformer electrical models based on their physical structure and the process technology parameters. Three-dimensional electromagnet-

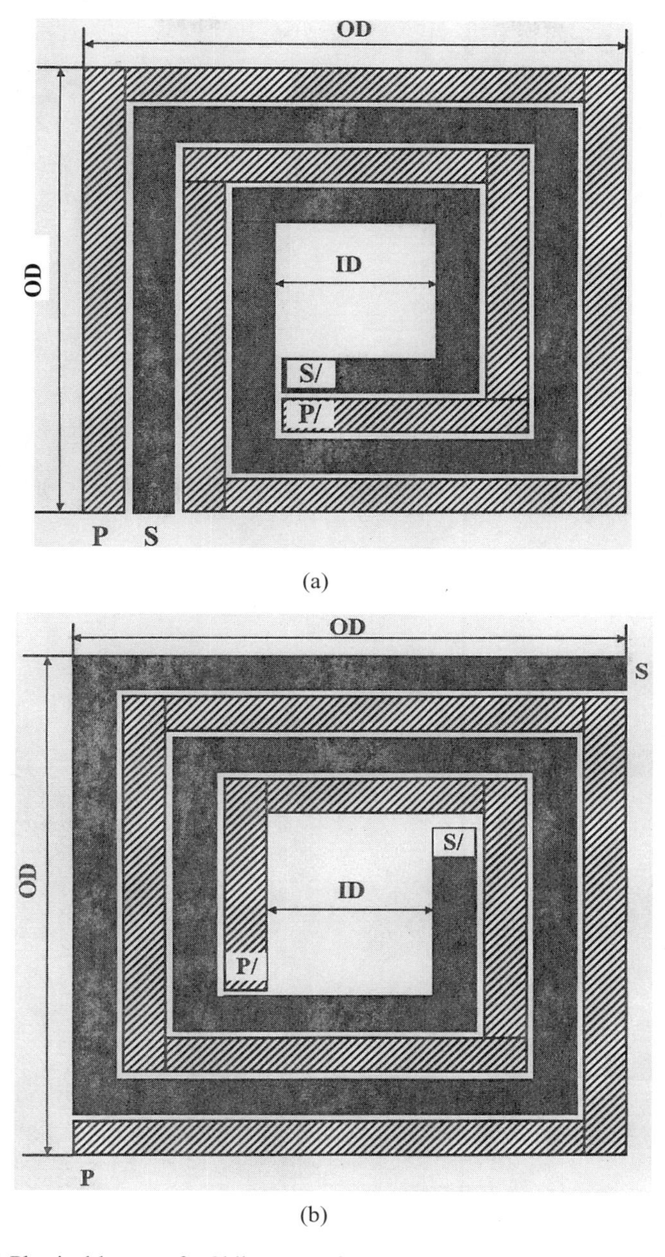


**Fig. 7.22.** (a) Physical layout of a Shibata transformer. (b) Physical layout of a Frlan transformer. (*Continued*)

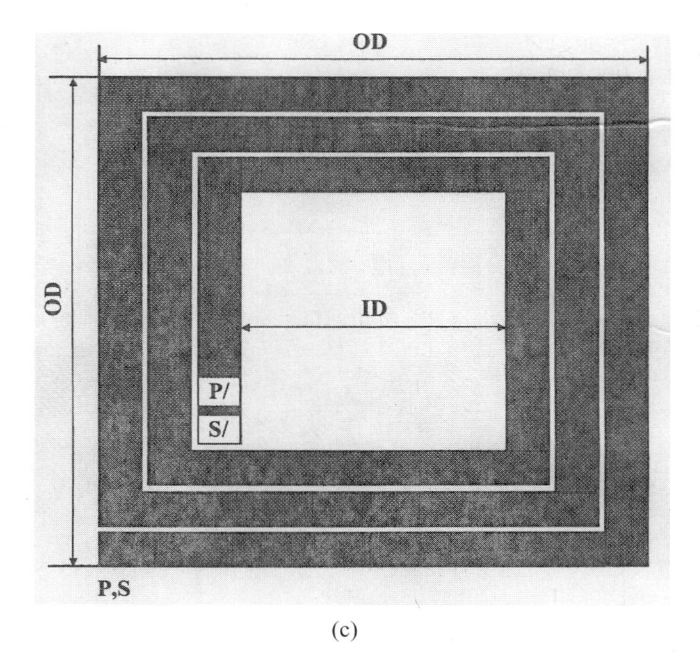


(c)

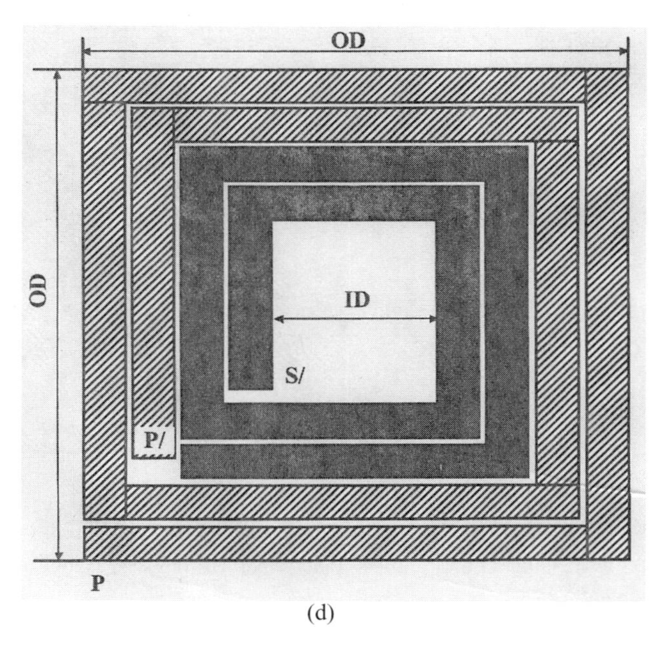


(d)

**Fig. 7.22.** (c) Physical layout of a Finlay transformer. (d) Physical layout of a single-turn coupling transformer.

ic simulators provide complete models for monolithic transformers but require significant simulation time.

Figure 7.23 illustrates a lumped-component equivalent structure for a transformer, in which the lumped element components can be obtained by the geometry and process technology parameters. Self-inductance and ohmic losses are represented by $L$ and $r(f)$, and the parasitic components are represented by $C_{ox}$, $C_{si}$, and $r_{si}$. Transformer action can be represented by mutual magnetic coupling, $M$, and electric coupling, $C_m$. In silicon-based processes, $r_{si}$ can model the losses caused by the electric field dissipation for $\rho > 1$ Ω-cm. The longitudinal currents in the substrate are induced by the individual currents flowing through the conductor segments, and the associated losses are proportional to square root of the frequency of operation.

Figure 7.24 shows a compact model for a transformer with four independently driven ports ($P$, $\overline{P}$, $S$, $\overline{S}$) and turns ratio ($1{:}n$). The transformer core is modeled by an ideal linear transformer with $L_m$ as magnetizing inductance and turns ratio of $1{:}n$. Inductors $L_P$ and $L_S$ placed in series with primary and secondary windings represent imperfect coupling and leakage of magnetic flux between the windings. Resistances $r_P$ and $r_S$ represent the ohmic losses in the windings. The interwinding capacitance is modeled by capacitors connected between the primary and secondary terminals and represented by $C_o$ and $C_x$. The parasitic components between each winding and the substrate are defined by $Z_{sh}$, and can be represented by the series connection of capacitors $C_{ox}$, and $C_{si}$, the substrate loss being represented by $r_{si}$, which is connected parallel to $C_{si}$.

Figure 7.25 shows several lumped-element models for hand calculations. These are directly obtained from theoretical equations. Mutual coupling between windings can be modeled by dependent current sources, and represented by $M_{21}$ and $M_{12}$, where $M_{21} = M_{12}$, for a reciprocal transformer. Thevenin equivalent of the first model is used to derive the second equivalent circuit. The turns ratio has been explicitly represented by the primary-to-secondary induced voltage, and the primary

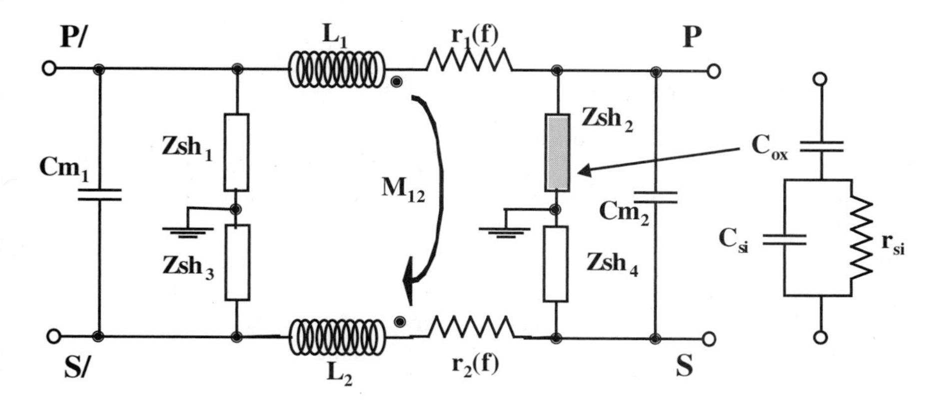


**Fig. 7.23.** Transformer model using lumped components.

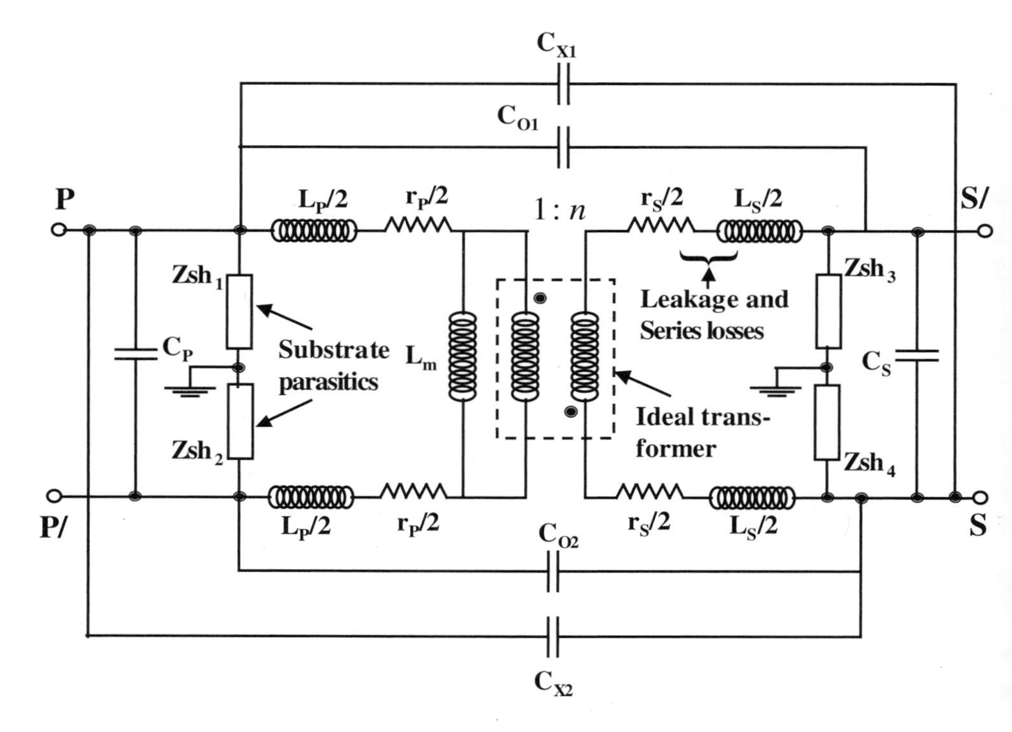


**Fig. 7.24.** Electrical model for a 1:*n* transformer.

voltage, $v_p$, is governed by the current, $i_p$, flowing through the primary winding inductance of $k_m L_P$. The terminal voltages are modified by the leakage inductances, $L_{kP}$ and $L_{kS}$, placed in series with the primary and secondary windings. The mutual inductance can be represented by $M^2 = (L_P - L_{kP})(L_S - L_{kS})$. The T-section model represents a very simple model to be used for hand calculation for IC design purposes.

### 7.5.4. Frequency Response of Transformers

A detailed analysis of the frequency response for a transformer is presented in [12]. The operational bandwidth of the transformer is determined by amplitude and phase imbalances between the two output terminals. Fig. 7.26 shows the lumped-element components for constructing the frequency response of transformers. Lumped components from the secondary terminal can be shifted to the primary terminal. The shunt inductor in the primary winding affects the lower cutoff in the frequency response by shunting energy to ground, whereas the series elements are responsible for the higher end of the frequency response by blocking the transmission of signal from primary to secondary as the frequency of operation increases.

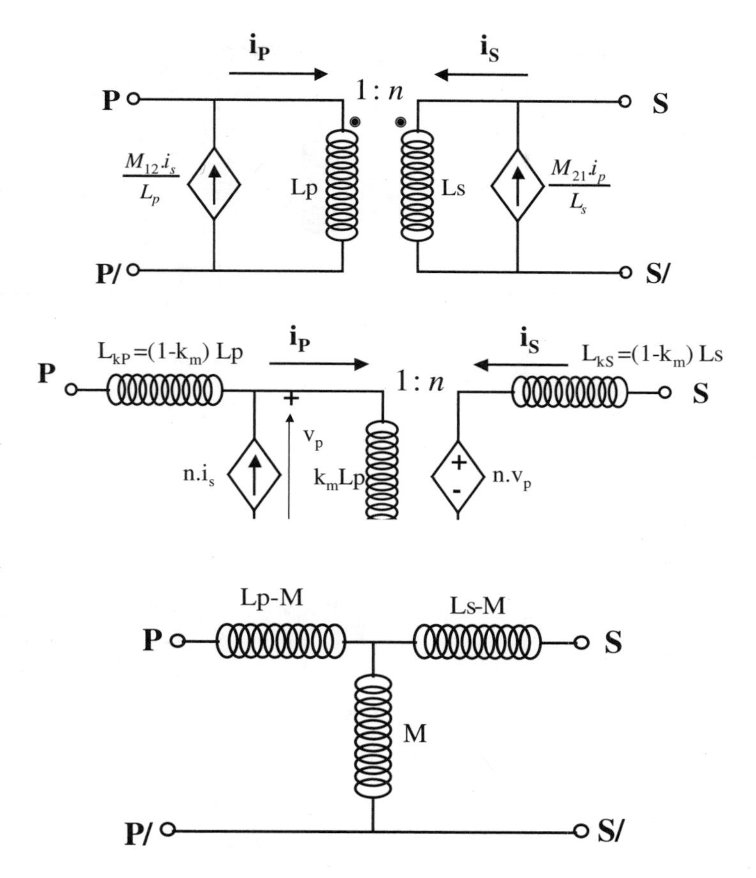


**Fig. 7.25.** Three types of transformer models for circuit analysis: direct form (top), dual source (middle), T-section model (bottom).

Frequency response at low frequency can be determined by transferring the secondary winding loss resistance, $r_S$, and the load resistance, $R_L$, to the primary side of the transformer as an equivalent resistance, $R'_P$, which is given by

$$R'_P = \frac{(R_L + r_S)}{(n/k_m)^2} \tag{7.8}$$

This resistance appears in parallel with the shunt magnetizing inductance, $L_m$. The DC value of the loss resistance, $r_S$, is determined from the total unwound length of the winding, width, and the metal sheet resistance, $r_{sh}$, and can be denoted by

$$r_{dc} = \frac{l_{winding}}{w} r_{sh} \tag{7.9}$$

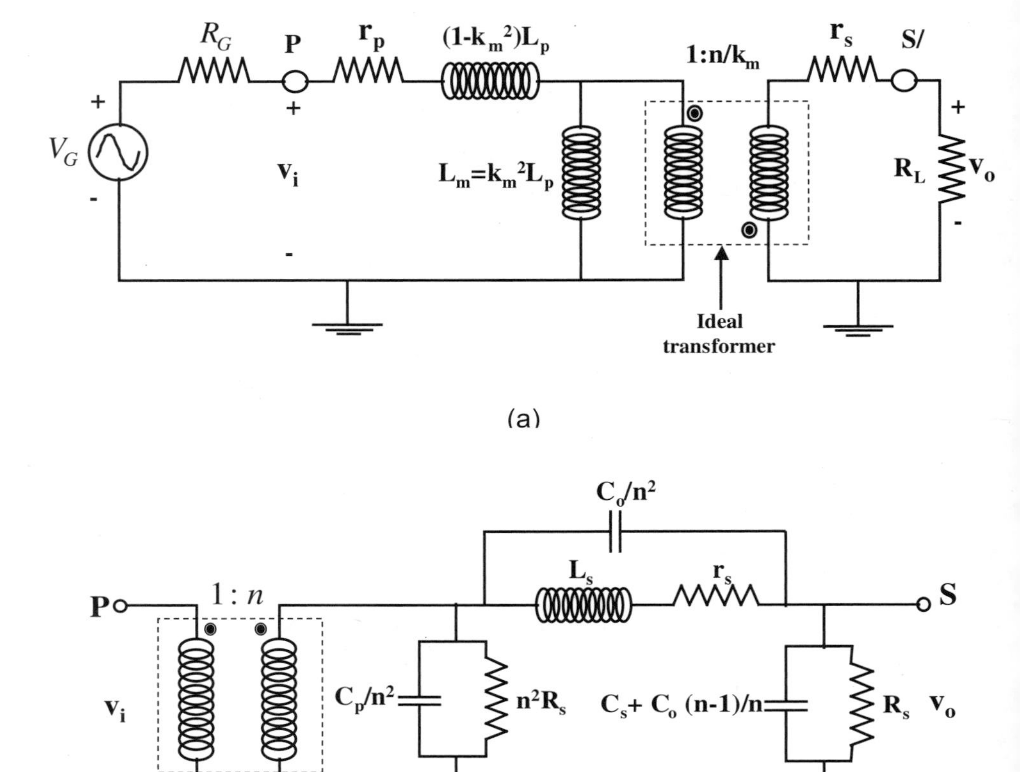


**Fig. 7.26.** (a) Low-frequency model with necessary components at the primary winding. (b) High-frequency model with necessary components at the secondary winding.

The unwound length of a square transformer with an integral number of turns, $N_t$, is given by

$$l_{\text{winding}} = 4N_t(OD - w) - 16(w + s)\frac{N_t(N_t - 1)}{2} \tag{7.10}$$

At fairly low operating frequencies, $\omega L_m$ is small compared to the equivalent primary resistance, $R'_P$, and very little of the input signal gets transferred to the secondary winding. With increase in the operating frequency, reactance of the magnetizing inductance becomes higher and its effect on the frequency response decreases. A simplified assumption can be made by assuming the reflected resistance at the primary side to be equal to the effective source resistance ($R_S + r_P = R'_P$) with an equal number of turns in the primary and secondary terminals, $n = 1$. The fractional bandwidth of the magnetic path is defined by

$$\frac{f_u}{f_l} = \frac{1 + k_m}{1 - k_m} \qquad \text{for } k_m < 1 \tag{7.11}$$

The fractional bandwidth increases with reduction of the $k_m$ factor. The transmission coefficient is given by $|S_{21}| = 2v_o/v_g$. The upper limit for the conversion loss of monolithic transformers can be set by a $k_m$ factor of unity, and ignoring the effects of ohmic loss in the winding and parasitic capacitances of the monolithic transformers, leading to a loss of 3 dB from primary to secondary, as $v_o/v_g = 1$.

With the increase of operating frequency, the series parasitics dominate, as the source and load impedances become much lower than the shunt parasitics. The lumped-element model shown in Fig. 7.26(a) could be used for interpreting the midband response with a simplistic assumption of a 1:1 turns ratio, with equal load on both windings, such that, $R = R_S + r_P = R_L + r_S$. The maximum signal transmission in the passband is given by [12]

$$|S_{21}| = \frac{2v_o}{v_g} = \frac{k_m R_L}{R} \tag{7.12}$$

This takes place at a peak frequency given by [12]

$$f_{peak} = \frac{R}{2\pi L_P \sqrt{1 - k_m^2}} \tag{7.13}$$

Thus, both the signal transmission and the fractional bandwidth decrease as $k_m$ decreases, and the minimum attenuation come closer to the lower cutoff frequency, $f_l = R/2\pi L_P$. A higher $k_m$ value improves the passband attenuation, as well as the usable bandwidth of the transformer.

The model shown in Fig. 7.26(b) can be used to interpret the high-frequency behavior of the monolithic transformer, with the necessary element shifted to the secondary. The turns ratio of 1: $n$ has been assumed with $k_m = 1$. The effect of the shunt inductance of the primary winding has been ignored at high frequency. The value of the interwinding capacitance is modified by the voltage transfer ratio, $n$, when it is transferred to the secondary terminal. For a noninverting connection, $n$ is positive, and the $\pi$ network at the secondary terminal of the transformer provides a bandpass response with a zero, leading to a notch in the high-frequency response. For an inverting connection, the value of $n$ is negative, leading to a positive reactance, which decreases with increasing frequency. This leads to a lowpass response filter with a higher cutoff frequency compared to that in case of the noninverting connection, implying a much higher bandwidth for these structures. A differential drive on the transformers improves the $Q$ factor of the monolithic transformer and leads to wider operating bandwidth with the same process technology.

### 7.5.5. Step-Up/Step-Down Transformer and Circuit Applications

In our earlier discussions, we have mostly focused on the 1:1 turns ratio of the transformer. However, step-up and step-down transformers are also used in monolithic circuits for impedance matching, especially at circuit interfaces where a circuit with high output impedance drives another circuit with low input impedance. In this subsection, we describe the formulation of higher turns ratios and the circuit applications of transformers in the developments of integrated RF front-ends.

A multiple turn ratio transformer can be developed by partitioning one winding (primary or secondary) into multiple individual turns, rather than one continuous winding. These multiple turns can be connected in parallel to form a step-up or step-down transformer, as needed. Figure 7.27 shows the schematic design of a 1:4 transformer. However, this layout can be transformed to a symmetric layout structure by using Rabjohn's structure, as described earlier.

A step-up transformer can be used as a narrowband alternative to a broadband resistive network. Feedback through the magnetic coupling in a transformer helps control amplifier gain and linearity, resulting in a much lower noise injection performance. A low supply voltage is also quite feasible as the ohmic drops are eliminated. Thus, step-up transformers can be thought of as ideal feedback elements in IC designs, as shown in Fig. 7.28. A symmetric structure of a step-up transformer and its equivalent circuit is shown in Fig. 7.29.

From a circuit design perspective, differential topologies are quite attractive for improved noise immunity. Thinking toward fully monolithic solution, the single-ended-to-differential functionality must be performed on-chip, and the two outputs must maintain a tolerable amount of amplitude and phase balance, for a tolerable

**Fig. 7.27.** A physical layout for a 1:4 transformer.

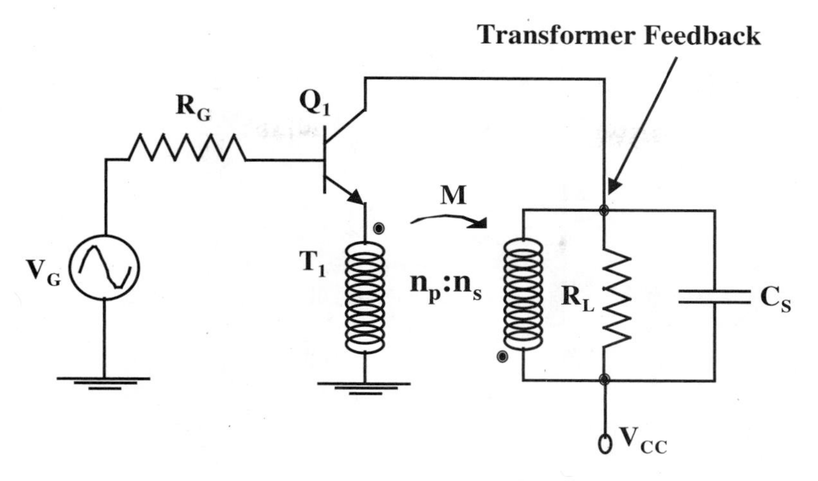


**Fig. 7.28.** Transformer-coupled feedback amplifier.

value of system SNR. Also, these circuits must provide a much lower value of signal distortion. Both of these design considerations are met by transformers, with the added advantage of zero DC power and lower noise contribution compared to their active counterparts. Wide bandwidth of transformers is especially helpful in development of wideband RF front-ends, which could be reused for different wireless applications. One potential problem with the transformer is the on-chip area requirement, and this is one reason why their placement must be carefully planned. Any circuit block with an inductive load at the output is a potential place for implementation of transformers to provide differential signals. One potential position for this is the LNA output, as shown in Fig. 7.30. Placing the transformer is also an attractive architectural decision, as the input impedance of different types of mixers vary widely with the selection of mixer topology. The provision for a center tap at the secondary winding of the transformer in the symmetric design also allows one to provide DC bias to the interfacing circuits.

## 7.6. ON-CHIP FILTERS

Continuing our mainstream discussion of fully monolithic implementations, we now focus on the fully monolithic implementation of front-end filters. As mentioned before, these filters are critical in rejecting out-of-band interferences. Traditionally, these filters were implemented off-chip, utilizing high-quality-factor passives. Implementation of such filters on-chip typically comes with a compromise on the front-end performances. Three approaches can be utilized for developing on-chip filters: (1) using passive components available in the technology, (2) using bond-wire inductance as a chip-package codesign, (3) using active filters at high-frequency regime. In this section, we discuss the latter two approaches.

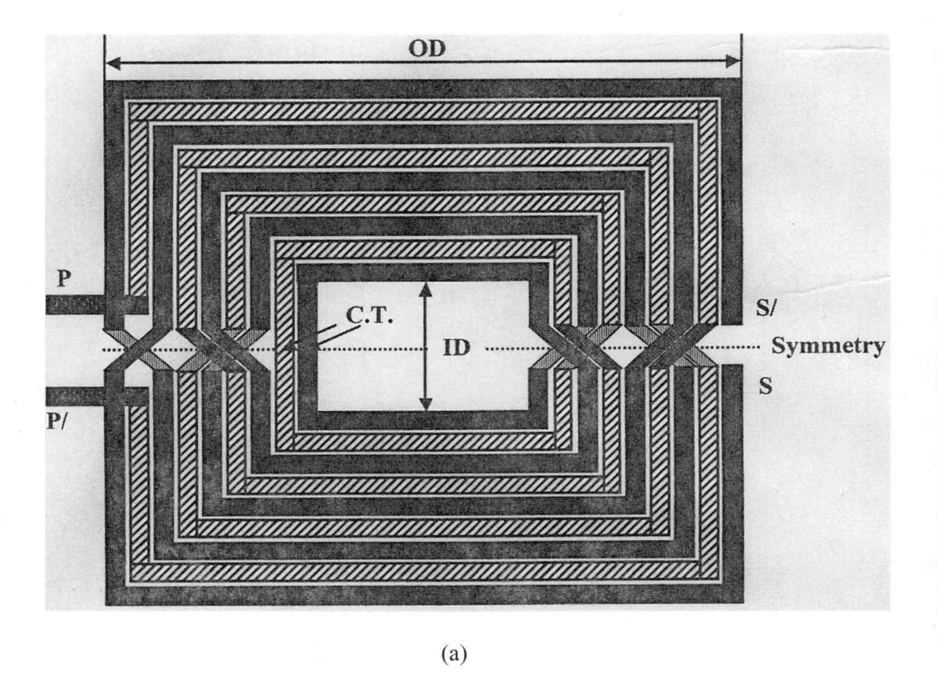


(a)

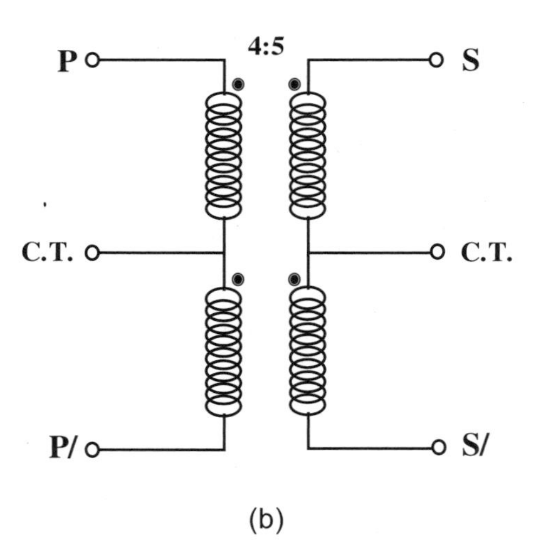


(b)

**Fig. 7.29.** (a) Physical layout. (b) Equivalent circuit.

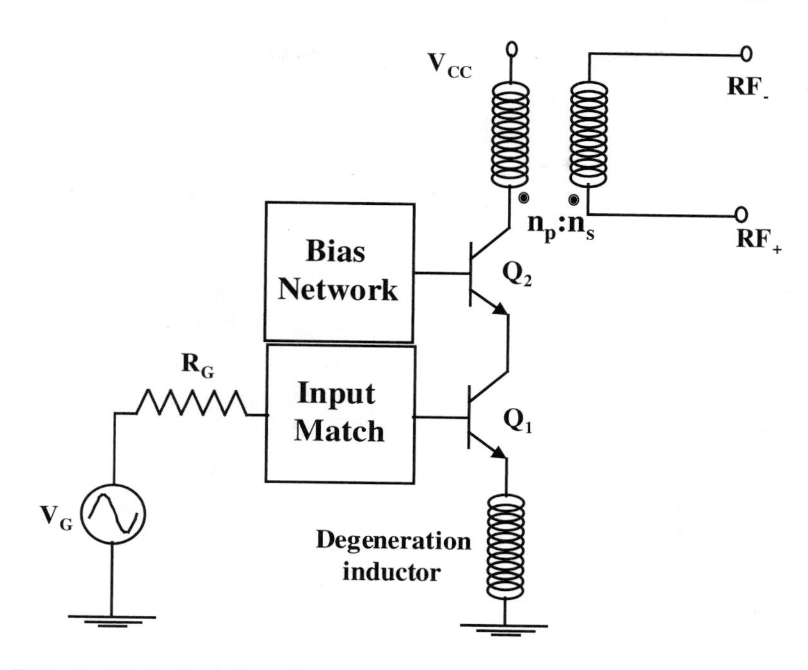


**Fig. 7.30.** An illustration of transformer load to generate differential signals for the mixer interface.

### 7.6.1. Filters Using Bond Wires

Bond wires can be utilized to develop on-chip filters [23]. The inductance of the bond wire can be utilized as a filter component. However, the bond-wire structure is not limited to a single bond-wire inductor, but could also include the inductance of a trace, which could be either a trace embedded in the IC carrier or a trace on the IC. Another configuration includes a bond wire to a pad on the IC or substrate connected to another pad in the IC, or a substrate connected back to another bond wire. Figure 7.31 shows the various types of bond-wire configurations available. Figure 7.32 shows the use of bond-wire components for development of RF front-end filters. The capacitors could be borrowed from standard semiconductor technologies.

### 7.6.2. Active Filters

A major challenge in front-end development is the reduction of chip size. A major part of the front-end area is consumed by inductors, which are commonly utilized for the purposes of tuning and matching. A common problem associated with close placement of multiple inductors in the front-end is their mutual coupling. In this subsection, we describe the active filter technique, which could be effectively utilized to eliminate passive inductor structures. It is interesting to note that active filters have been used in conventional analog designs for a long time [24]. The basic

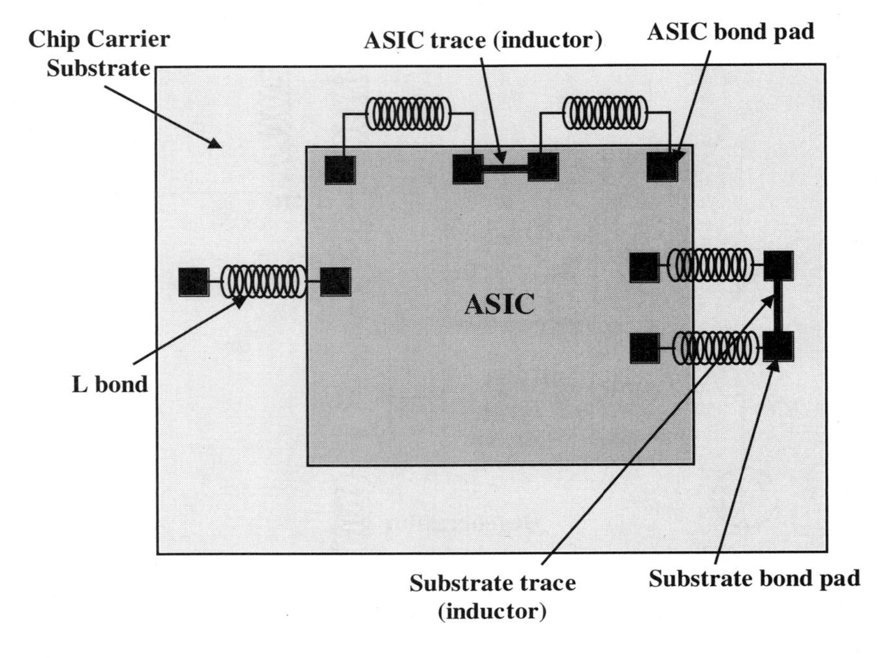


**Fig. 7.31.** Availability of various bond-wire structures.

concept of active filters starts from the realization of inductors by using an antiparallel connection of transconductor blocks, with a capacitor connected as the load. A basic gyrator core is shown in Fig. 7.33.

A capacitor in conjunction with a gyrator core can be utilized in realizing an active inductor. This leads to space compaction compared to the passive counterpart.

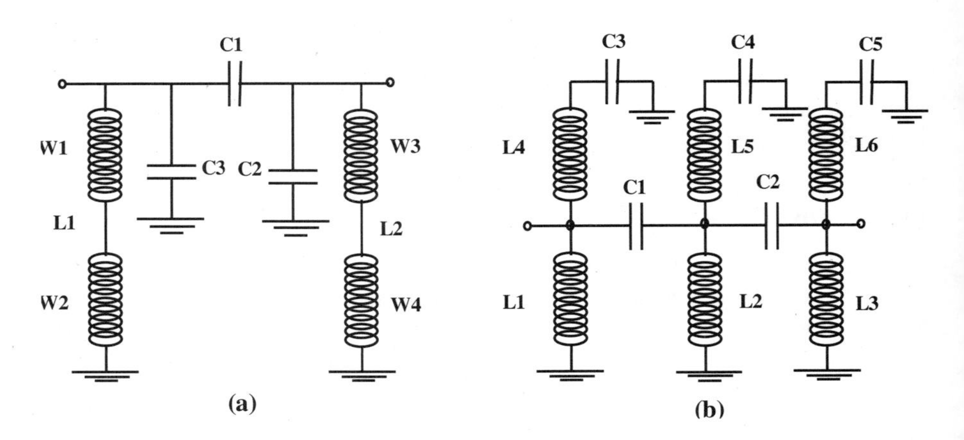


**Fig. 7.32.** Filter structures using bond wires.

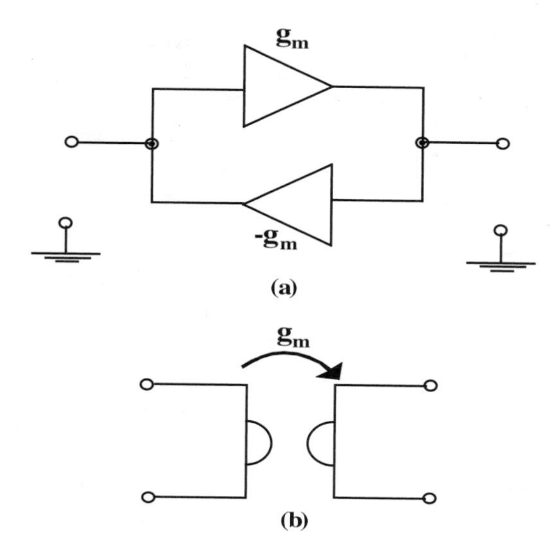


**Fig. 7.33.** A basic gyrator core and its symbol.

This concept can be further extended to implement an L–C tank circuitry. A gyrator structure can be realized in IC technologies by connecting two transconductance cells in negative feedback fashion, as shown in Fig. 7.33. The maximum achievable DC voltage gain ($A_{DC} = g_m/g_{out}$) is determined by the transconductance of each cell and the output transconductance ($g_{out}$). The equivalent circuit at the input of the gyrator terminated in a capacitance at the other port consists of an inductor, $L$, with series loss resistance, $R_s$, and shunt resistance, $R_p$, a shown in Fig. 7.34. The values of the equivalent circuit components are denoted by [24]

$$L = \frac{C}{g_m^2}, \qquad R_s = \frac{g_{out}}{g_m^2}, \qquad R_p = \frac{1}{g_{out}} \tag{7.14}$$

An L–C tank can be obtained by attaching a capacitor, $C$, at both ports of the gyrator network. The susceptances of the capacitor and the inductor are each $g_m$ at the resonant frequency, given by $\omega_0 = g_m/C$, and the $Q$ factor of the active inductor can be obtained by using a narrow-band approximation to transform $R_p$ into a series resistance, $R_s' \approx (\omega_0 L)^2/R_p = g_{out}/g_m^2$, such that

$$Q = \frac{\omega_0 L}{R_s + R_s'} = \frac{g_m}{2g_{out}} \tag{7.15}$$

In this way, the $Q$ factor is half of the DC gain of the transconductors used in the gyrators, independent of the resonance frequency.

Having discussed the basic principles of active filters, we now describe three important effects associated with the performance of active filters. The first in this cat-

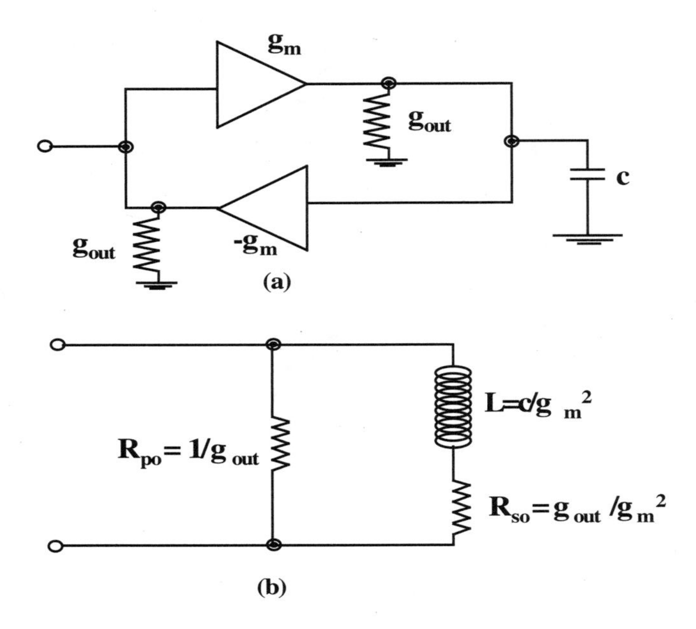


**Fig 7.34.** (a) Gyrator terminated in a capacitor. (b) Equivalent circuit of lossy inductor.

egory is the effect of high-frequency phase shifts. The internal poles associated with each of the transconductance cells produce produce phase shifts at high frequencies that result in undesirable frequency response as the operating frequency approaches these pole frequencies. When the two transconductor blocks present in the gyrator have a phase lag (represented by $e^{-j\phi}$), the terminal admittance with a capacitor connected at the other port is denoted by [24]

$$Y_{\text{in}} = \frac{g_{\text{m}}^2 e^{-j\phi}}{j\omega C} \approx \frac{g_{\text{m}}^2}{j\omega C} - \frac{2g_{\text{m}}^2}{\omega C}\phi \tag{7.16}$$

With the assumption that the gyrator has one dominant high-frequency pole ($\omega_1$), the phase shift can be approximated by $\phi = \tan^{-1}(\omega/\omega_1) \approx \omega/\omega_1$, leading to

$$Y_{\text{in}} \approx \frac{1}{j\omega L} - \frac{2}{\omega_1 L} \tag{7.17}$$

This effect can be taken into consideration by connecting a negative resistance $-R_{\text{p1}}$ in parallel with the active inductor, as shown in Fig. 7.35. This parallel resistance, $R_{\text{p}}$, determines the passband gain at resonant frequency $f_0$ when the LC resonant tank is driven by a current source [24].

It is important to evaluate active filters for their noise contribution. Noise in a resonator built from a physical inductor and capacitor arises from the dissipation in the individual elements. Noise is contributed by inductor loss, capacitor leakage,

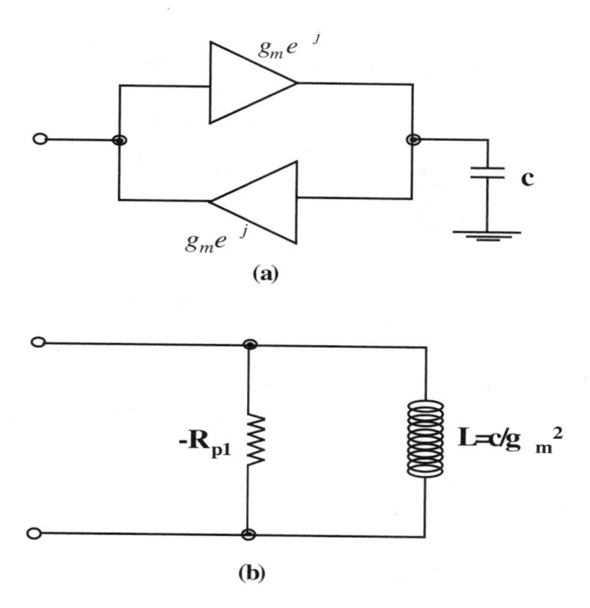


**Fig. 7.35.** (a) Phase lag in the gyration transconductors. (b) Negative loss in the simulated inductor.

and external resistances in conjunction with the reactive elements. The r.m.s. broadband noise voltage of $(kT/C)^{1/2}$ appears across the terminals of a parallel LC resonator with any positive dissipation. Noise is a major limitation for active filter structures, and is usually subsequently higher than that from their passive counterparts. Fig. 7.36 shows the noise sources available in a gyrator. As the output conductance of the transconductors is small, the noise sources can be put outside the gyrator loop. This gives rise to the equivalent circuit as shown in Fig. 7.36(b), with an inductor with shunt noise current $i_{\mathrm{nL}} = g_{\mathrm{m}} v_{\mathrm{n}}$ and a series noise voltage $v_{\mathrm{nL}} = v_{\mathrm{n}}$. The noise spectral density across the terminals of an LC resonant circuit at resonance, constructed with equal capacitances connected at the two ports of the gyrator and loaded with a large resistance $R$, is given by [24]

$$\hat{v}_{\mathrm{n}}^2 = \hat{i}_{\mathrm{nL}}^2 R^2 + \hat{v}_{\mathrm{nL}}^2 \left( \frac{R}{\omega_0 L} \right)^2 = 2(g_{\mathrm{m}} R)^2 \, \hat{v}_{\mathrm{n}}^2 \tag{7.18}$$

Assuming the input stage of a gyrator as a MOS differential pair, $\hat{v}_{\mathrm{n}}^2 \approx kT/g_{\mathrm{m}}$, the noise bandwidth for a high-$Q$ resonator to be approximately $f_0/Q$, and using the relationship $R = Q/\omega_0 C = Q/g_{\mathrm{m}}$, the r.m.s. noise voltage across the resonator can be given by

$$v_{\mathrm{n}} \sim \left[ (g_{\mathrm{m}} R)^2 \left( \frac{kT}{g_{\mathrm{m}}} \right) \left( \frac{\omega_0}{Q} \right) \right]^{1/2} = \sqrt{Q} (kT/C)^{1/2} \tag{7.19}$$

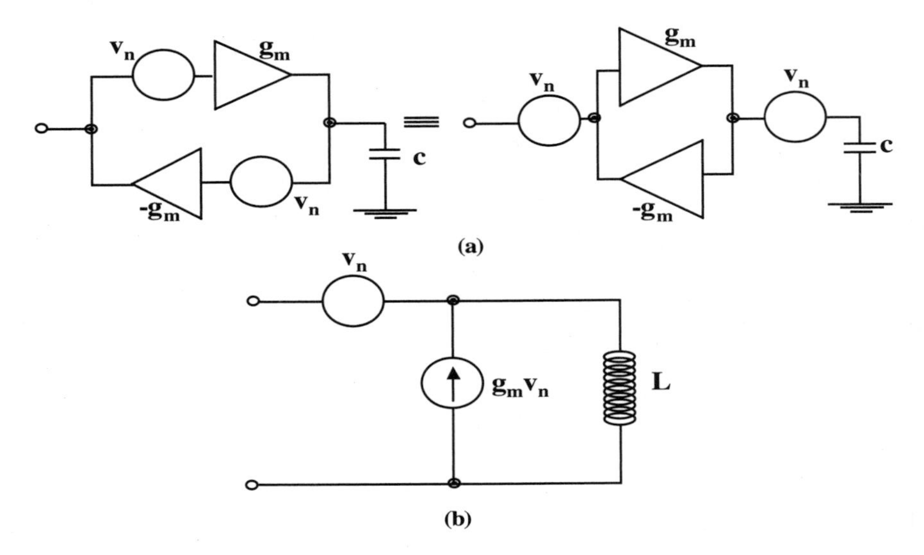


**Fig. 7.36.** (a) Effect of noise in transconductors. (b) Noisy simulated inductor.

This shows that the noise power for active filters is $Q$ times worse than their passive counterparts, and this is a potential drawback for high-performance transceivers. Figure 7.37 shows a representation of a bandpass filter with a noisy simulated inductor.

It is noteworthy that the active filters are also quite limited in their dynamic range performance. A detailed illustration of dynamic range is presented in [21], and in compact form, this can be presented as

$$DR = \alpha V_{\mathrm{DD}} \sqrt{\frac{C}{kT}} \sqrt{\frac{1}{Q}} \tag{7.20}$$

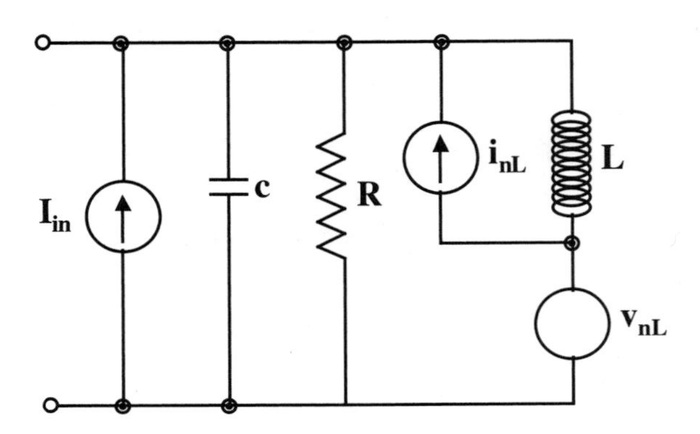


**Fig. 7.37.** Bandpass filter representation with a noisy simulated inductor.

where $\alpha$ is the fraction of supply voltage $V_{DD}$ used for considering distortion. Thus, the dynamic range worsens with increasing $Q$ of the resonator, and can be held constant if the parameters of Eqn. 7.20 can be made proportional to each other, such as

$$C \propto Q, \qquad g_{\mathrm{m}} = \omega_0 C \propto \omega_0 Q, \qquad \Rightarrow A \propto \omega_0 Q \tag{7.21}$$

where denotes the chip area, and it has been assumed that $g_{\mathrm{m}} \propto I_D$ at a fixed $(V_{\mathrm{GS}} - V_{\mathrm{t}})$, a somewhat reasonable approximation for short-channel MOSFETs.

## 7.7. ON-WAFER ANTENNA

In the previous sections, we have described antennae on substrates other than silicon. Implementation of antennae in silicon wafers is quite attractive for more integration on-chip. However, several major obstacles must be overcome for realization of practical on-wafer antennae: (a) very large dimension at commercial wireless bands, (b) lossy nature of silicon substrates, (c) dissipation of radiation pattern in silicon, (d) isolation from other parts of the integrated transceiver. Despite these obstacles, attempts have been made to develop antennae in silicon-based substrates [23–24]. However, these use high-resistivity silicon substrates for their implementation and sometimes resort to sophisticated techniques such as micromachining [23].

Microstrip-patch and slot-based topologies are quite popular for development of antennas in the millimeter wave frequency range. Slotline antennae provide the advantage of integration with active devices in a coplanar manner, thus eliminating the need for via holes. This antenna topology also provides an interesting option of developing an antenna array for combining high-frequency signals from many devices.

We now focus on two reports of on-wafer silicon antennae using high-resistivity silicon substrates. Silicon has a high dielectric constant of 11.9, and when used as a planar antenna substrate results in small antenna size, narrow bandwidth, and low radiation efficiency [23]. Development of an antenna using a reverse-side etching process is demonstrated in [23]. The substrate edges are made to taper away from transmission line conductor edges, thus leaving an overhanging conductor, and giving rise to trenches along the radiation edges as shown in Fig. 7.38. Figure 7.39 shows the localized fields occurring as a result of the reverse-side etching process for various etching methodologies.

An interesting set of applications and circuit/system integration effects can be developed using on-wafer antennas. One such example has been shown in [24]. The substrate used in this application is a high-resistivity silicon (HRS) without an insulating barrier underneath the upper conductor. The patch pattern has been developed on the polished side of the antenna, with the ground plane on the etched side. Thus, a metal–semiconductor barrier is formed, which gives rise to a distributed Schottky barrier diode connected from the antenna feed point to the ground. This configuration is shown in Fig. 7.40. The antenna, along with the diode, gives rise to a direct signal demodulation, which could be used to detect modulated signals. The

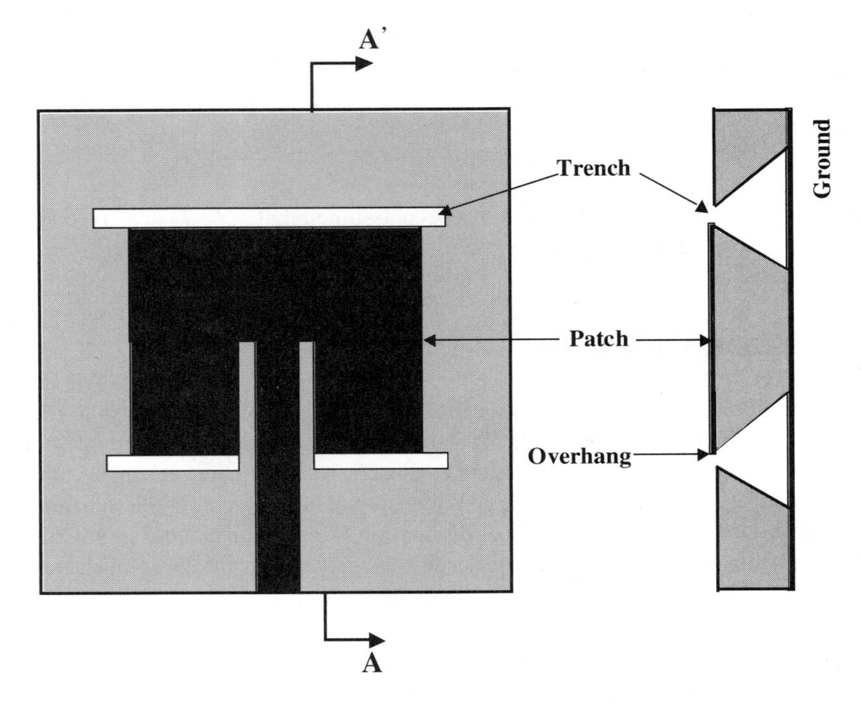


**Fig. 7.38.** Illustration of a reverse-side etched patch antenna.

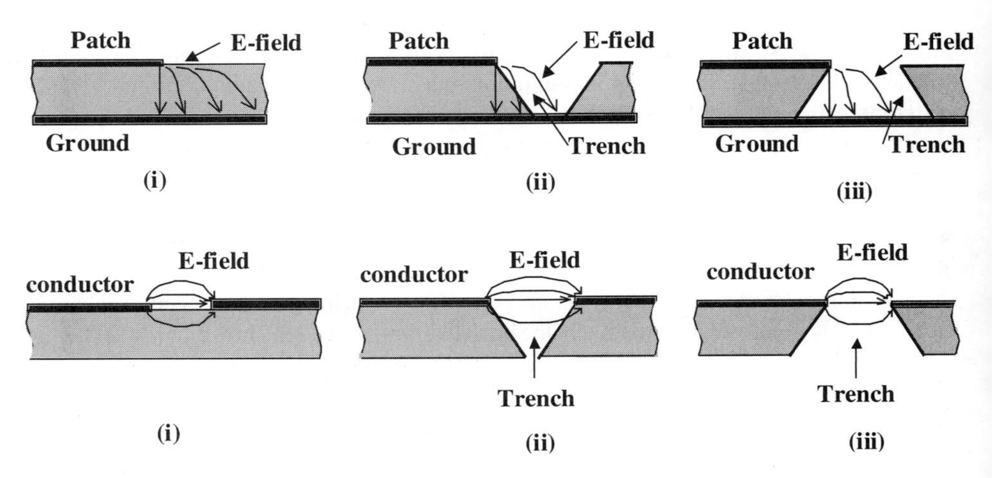


**Fig. 7.39.** Electric fields produced by (a) no trench, (b) front-side etching, (c) reverse-side etching.

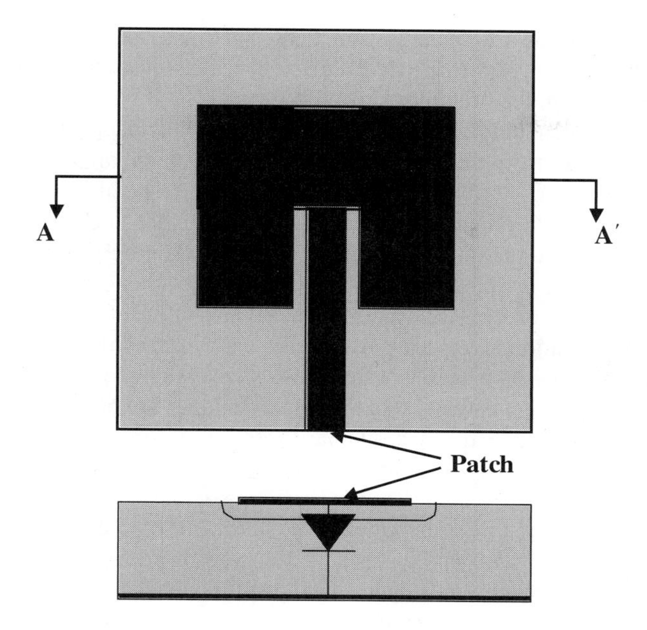


**Fig. 7.40.** Illustration of a patch antenna direct modulation system.

diode characteristics change with the applied DC bias, which can change the impedance of the antenna–diode combination. One interesting aspect of this approach is the controllability of the antenna parameters with the application of DC bias. The DC bias influences the antenna input match, resonant frequency, and the radiation pattern. This could be quite suitable for a wideband operation. The bandwidth of the detected signal is limited by two effects: (1) the diode turnoff period limitation, mainly caused by the high-resistivity substrate used for antenna construction; (2) parasitic capacitance of the large volume of the distributed diode.

## 7.8. WAFER-LEVEL PACKAGING

Having discussed various types of passive components, we now focus on the packaging aspects for realization of a fully integrated front-end. Package development for wireless radios is dominated by the requirement to create smaller, thinner, and lighter solutions with lower cost. Packaging of ICs involves a variety of assembly and packaging steps. The success of an effective packaging scheme depends heavily on the package cost, its impact on the circuit and system performances, and its reliability. Next-generation packages that provide low cost include stacked packages such as lead-frame-based chip-scale packages (CSPs) such as Amkor's MLF, QON, and SONs, and Fujitsu's BCC. These packages are usually targeted for low I/O count, do not provide solder balls, and are quite inexpensive. A typical wireless solution with multiple ICs contain several CSPs such as baseband processor, DSP, SRAM, flash memory, and EEPROM. Different wafer-level packaging techniques have been reported in the literature [25–26].

The continued demand for integration and smaller form factor for wireless solutions has given rise to the development of wafer-level packaging. In the case of wafer-level packaging, the whole assembly takes place on the wafer, leading to lower cost when batch processing is performed. These are packaged and tested at the wafer level before the dicing operation. In this packaging process, contacts are rewired on top of the wafer and ICs and thin passive components are assembled by flip-chip technology or thin-film interconnect. Deposition of polymer passivation and stress compensation layers are performed locally, and solder balls are placed to realize small systems on a wafer. After these steps, wafer-level testing is carried out, and then the product is marketed. Similar techniques are often used in the packaging processes of MEMS devices, such as pressure sensors, etc. These techniques will become more and more popular in the future generation of wireless communication technologies.

Wafer-level packaging can be categorized into four classes: (1) redistribution, (2) copper post/bump with epoxy resin, (3) encapsulated bond, (4) encapsulated beam lead. The redistribution category includes many packages, which are considered flip chips. A copper post or bump could be used if protective encapsulation is provided by epoxy resin and a solder ball is attached. It should be noted that in a system-in-package (SiP) approach, the ICs and passive components are packaged together in a functional module, and the packages may have a higher I/O cost than a single-chip package, but might provide a cost saving from a system perspective. Although system-on-chip (SoC) is useful for most integrated systems, it might not be appropriate for most complex mixed-signal designs.

## 7.9. CONCLUSION

In this chapter, we have discussed various passive integration techniques, system-on-package, and system-on-chip integration techniques, and described some of the emerging applications. Both of the integration techniques, SoP and SoC, provide great potential for miniaturization in the RF front-end. It must be noted that various forms of SoP solutions are already in place for most of the wireless applications. Although there are major research activities in terms of realizing on-chip passives, it still remains to be seen whether a fully integrated solution will become a reality in the near future.

## REFERENCES

1. S. Donnay, P. Pieters, K. Vaesen, W. Diels, P. Wambacq, W. D. Raedt, E. Beyne, M. Engels, and I. Bolsens, "Chip-package codesign of a low-power 5-6GHz RF front end," *Proceedings of the IEEE, 88,* 10, October 2000.

2. A. Sutono, A.-V. Pham, J. Laskar, and W. R. Smith, "RF/microwave characterization of multilayer ceramic-based MCM technology," *IEEE Transactions on Advanced Packaging, 22,* 3, 326–331, August 1999.
3. R. C. Frye, "MCM-D implementation of passive RF components: Chip/package tradeoffs," in *IEEE Symposium on IC/Package Design Integration,* February 1998, pp. 100–104.
4. A. Iqbal, M. Swaminathan, M. Nealon, and A. Omer, "Design tradeoffs among MCM-C, MCM-D and MCM-D/C technologies," in *IEEE Multichip Module Conference,* March 1993, pp. 12–17.
5. A. Sutono, J. Laskar, and W. R. Smith, "Development of integrated three-dimensional Bluetooth image reject filter," *IEEE International Microwave Symposium Digest, 1,* 339–342, June 2000.
6. S. Pinel, S. Chakraborty, M. Roellig, R. Kunze, S. Mandal, H. Liang, C.-H. Lee, R. Li, K. Lim, G. White, M. Tantzeris, and J. Laskar, "3D Integrated LTCC module using μBGA technology for compact C-band front-end module," *IEEE International Microwave Symposium Digest, 1,* 1553–1556, June 2002.
7. J. R. Long and M. A. Copeland, "The modeling, characterization, and design of monolithic inductors for silicon RF IC's," *IEEE Journal of Solid State Circuits, 32,* 3, 357–369, March 1997.
8. F. W. Grover, *Inductance Calculations.* Princeton, NJ: Van Nostrand, 1946, reprinted by Dover Publications, New York, 1954.
9. D. Kammler, "Calculation of characteristic admittances and coupling coefficients for strip transmission lines," *IEEE Transactions of Microwave Theory and Techniques, MTT-16,* 925–937, November 1968.
10. E. Pettenpaul, H. Kapusta, A. Weisgerber, H. Mampe, J. Luginsland, and I. Wolff, "CAD models of lumped elements on GaAs up to 18 GHz," *IEEE Transactions of Microwave Theory and Techniques, MTT-36,* 294–304, February 1988.
11. M. Danesh and J. R. Long, "Differentially driven symmetric microstrip inductors," *IEEE Transactions on Microwave Theory and Techniques, 50,* 1, 332–341, January 2002.
12. J. R. Long, "Monolithic transformers for silicon RF IC design," *IEEE Journal of Solid State Circuits, 35,* 9, 1368–1382, September 2000.
13. K. Shibata, K. Hatori, Y. Tokumitsu, and H. Komizo, "Microstrip spiral directional coupler," *IEEE Transactions of Microwave Theory and Techniques, 29,* 680–689, July 1981.
14. E. Frlan, S. Meszaros, M. Cuhaci, and J. Wight, "Computer-aided design of square spiral transformers and inductors," *IEEE International Microwave Symposium Digest,* June 1989, pp. 661–664.
15. H. J. Finlay, "U.K. patent application," 1985, 8 800 115.
16. J. N. Burghartz, D. C. Edelstein, M. Soyuer, H. A. Ainspan, and K. A. Jenkins, "RF circuit design aspects of spiral inductors on silicon," *IEEE Journal of Solid State Circuits, 33,* 12, 2028–2034, December 1998.
17. J. N. Burghartz, M. Soyuer, K. A. Jenkins, M. Kies, M. Dolan, K. J. Stein, J. Malinowski, and D. L. Harame, "Integrated RF components in a SiGe bipolar technology," *IEEE Journal of Solid State Circuits, 32,* 9, 1440–1445, September 1997.
18. T. H. Lee, "CMOS RF: No longer an oxymoron," in *Technical Digest, GaAs IC Symposium,* 1997, pp. 244–247.

19. H. Samavati, A. Hajimiri, A. Shahani, G. N. Nasserbakht, and T. H. Lee, "Fractal capacitors," *IEEE Journal of Solid State Circuits, 33,* 12, 2035–2041, December 1998.
20. M. A. I. Mostafa, J. Schlang, and S. Lazar, "On-chip RF filters using bond wire inductors," *Proceedings of IEEE International ASIC/SOC Conference,* 2001, pp. 98–102.
21. Y.-T. Wang and A. A. Abidi, "CMOS active filters at very high frequencies," *IEEE Journal of Solid State Circuits, 25,* 6, 1562–1574, December 1990.
22. C. Guo, C.-W. Lo, Y.-W. Choi, I. Hsu, T. Kan, D. Leung, A. Chan, and H. C. Luong, "A fully integrated 900-MHz CMOS wireless receiver with on-chip RF and IF filters and 79-dB image rejection," *IEEE Journal of Solid State Circuits, 37,* 8, 1084–1089, August 2002.
23. Q. Chen, V. F. Fusco, M. Zheng, and P. S. Hall, "Micromachined silicon antennas," in Proceedings of IEEE international conference on microwave and milimeter wave technology, pp. 289–292, 1998.
24. V. F. Fusco and Q. Chen, "Direct signal modulation using a silicon microstrip patch antenna," *IEEE Transactions on Antennas and Propagation, 47,* 6, 1025–1028, June 1999.
25. E. J. Vardaman, "Low cost options for next generation packaging," in *IEEE Electronics Packaging Technology Conference,* 2002, pp. 457–459
26. R. Islam, C. Brubaker, P. Lindner, and C. Schaefer, "Wafer level packaging and 3D interconnect for IC technology," in *IEEE/SEMI Advanced Semiconductor Manufacturing Conference,* 2002.

# 8

# DESIGN FOR INTEGRATION

## INTRODUCTION

In this chapter, we describe some of the issues concerning the design of a multifunction IC. Complex ICs must be viewed as subsystems instead of individual components. By including multiple functions spanning various design disciplines, both the complexity and the number of elements increase, and planning for integration becomes a necessity. In this chapter, we highlight the many challenges unique to an integrated IC solution and provide insight and guidelines on how to address them.

## 8.1. SYSTEM DESIGN CONSIDERATIONS

When designing an IC with a high level of integration of RF and analog and digital circuitry, many challenges have to be overcome for a successful integration. Each type of circuitry has very specific characteristics and requirements that are difficult to address simultaneously. In this chapter, we examine the major issues concerning the integration and expand on integration in silicon-based semiconductor technology.

### 8.1.1. I/O Count

One of the major issues with multifunction ICs is the difficulty of routing input and output (I/O) signals. When integrating multiple individual integrated circuits (ICs) in a single chip, a great number of I/Os from every individual IC will have to be consolidated in a single IC. Due to the lack of periphery and presence of limited


ISBN 0-471-22591-6 

number of bonding locations and pins in a given package, it becomes very difficult to maintain the same number of I/Os when integrating multifunction ICs into a single IC. Some I/Os will turn into internal interconnections by the virtue of integration, but many will have to go off-chip for filtering or use of other off-chip components such as large capacitors and inductors. In general, the I/O density increases with integration and creates great difficulties for packaging and IC floor planning.

Ball grid array (BGA) and flip-chip packages allow for increase in interconnect density but they come with their own set of problems. There are parasitics associated with the internal routing of the package that can cause problems at high frequencies. Use of such packages is easier for lower-frequency functions, but as the frequency of operation increases, the added parasitics of such packages become more and more dominant, making them less feasible for use at higher frequencies. At RF frequencies beyond a few gigahertz, the performance degradation caused by the high-density interconnects can work as a good deterrent for using these types of packages.

One method for addressing the issue of interconnect density is to use receiver architectures that simplify the receiver lineup. One such architecture is direct conversion. By reducing the number of off-chip filters and the entire IF chain, direct conversion allows for reduction in the number I/Os required for a receiver.

### 8.1.2. Cross-Talk

As the level of integration increases, the number and density of signal lines that travel in and out of the IC and the package also increases. This reduces the physical distance between the signal paths, and increases coupling and unwanted interaction between various signals. In the case of multifunction ICs, in which digital, analog, and RF functions and circuitry are integrated on the same die, the problem of cross-talk can be quite significant. Clock and control signals that are used in the digital portions of the IC can interact with analog and RF portions of the circuit and create interference and noise, which are not observed when these functions are implemented in multiple ICs. This is a great challenge that will have to be addressed in any IC with a high level of integration.

In addition, the impact of package layout and orientation of wire bonds should not be forgotten. Careful allocation of package pins can greatly help to reduce the possibility of cross-talk between pins and the bond wires connecting the package pins to the bond pads on the die. Bond wires should be treated as potential antennas at higher frequencies and careful EM simulations must be performed to estimate the necessary isolation between critical bond wires. Placing bond wires in perpendicular orientation can often help to reduce their coupling with one another dramatically. As shown in Fig. 8.1, this can be explained by observing the magnetic field surrounding two wires. By placing the wires perpendicular to one another, the EM fields associated with each wire never cross each other, creating no interfering currents.

An efficient method for reducing crosstalk is the use of differential topology. In a differential circuit, an existing interferer becomes a common-mode signal that can be easily rejected by the common-mode rejection characteristic of a design.

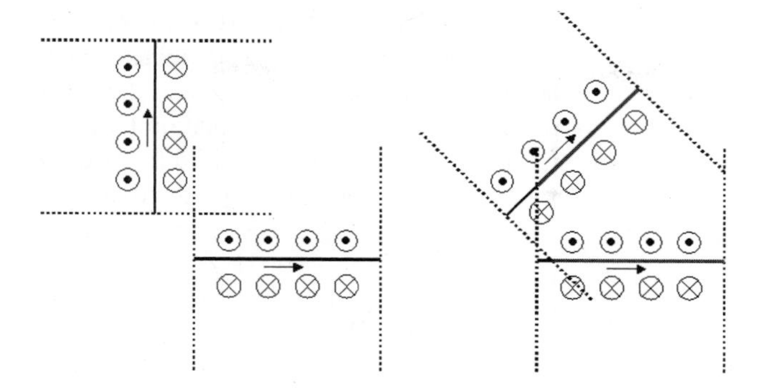

**Fig. 8.1.** Interaction between two adjacent wire bonds.

### 8.1.3. Digital Circuitry Noise

Digital noise is generated from the harmonics of the clocks and other signals used in the digital circuitry. As shown in Eqn. 8.1, the nature of a digital signal dictates a broadband signal consisting of odd harmonics of the fundamental frequency of the square wave. These harmonics spread into the higher frequencies and can act as noise, contaminating the analog and RF signal paths in the IC.

$$V(t) = A\sin(wt) + B\sin(3wt) + C\sin(5wt) + \ldots \tag{8.1}$$

Digital noise can penetrate other circuitry through the substrate or any common supply or ground connection. This makes it very essential to isolate the digital circuitry with isolation barriers and separate the supply and ground connections as much as possible. This typically requires the use of large off-chip capacitors to bypass the supply lines and provide a good AC ground.

As mentioned before, use of differential circuitry can help with the problem of digital noise. However, a combination of various techniques must be employed to obtain the adequate rejection required for the sensitivity levels of most modern high-performance receivers.

## 8.2. IC FLOOR PLAN

IC floor planning is a critical step in implementing an IC with multiple functions and circuitry. This involves careful allocation of each individual block of the receiver and preplanning for signal flow through the board, package, and die. Although the major portion of this process only involves the die and package, the impact of the external board circuitry that surrounds the chip should also be taken into consideration.

IC floor planning can be very crucial in reducing cross-talk between various circuits. Judicious allocation of die area and location to various functional circuits can help reduce the problem of cross-talk. Isolation is a function of substrate resistivity, distance, and isolation barriers that are implemented as shields around various circuits. For the case of a silicon substrate, in which the substrate conductivity is high, isolation barriers can help to improve isolation by increasing the resistivity around a circuit and act as barriers to signal interference. Deep trenches and grounded substrate contact rings have proven very effective in accomplishing this job.

### 8.2.1. Signal Flow and Substrate Coupling

Signal flow is determined by an external requirement for input, output, supply, and control pin locations. The chip will have to be designed for easy interaction with the other components for the development of a wireless product. Therefore, the signal flow inside the chip will have to be compatible with the signal flow in the entire system. Signal flow will also be impacted by the need for isolation between various components or signals in the receiver. For example, when routing the signal, one should always consider the isolation between LO and RF and IF signals so that the receiver is not affected by interaction of the LO signal with other components of the receiver.

Routing of the signal inside the chip has direct impact on the routing on the board on which the receiver IC is mounted. A poor arrangement of the critical signal path can cause many problems. For example, if the LO signal is supplied from a separate synthesizer, and the LO port is placed at an inconvenient location, the board trace that connects the synthesizer to the LO port may have to cross over other traces carrying RF or supply lines. This could result in LO leakage into the RF lines or the supply line, causing various problems such as LO self-mixing or amplitude modulation.

Other issues that are internal to each circuit block should also be considered in determining the signal flow inside each block. Feedback is one of the major concerns while dealing with the internal structure of each individual block. Negative feedback impacts the system by giving rise to poor amplifier gain performance, whereas positive feedback causes unwanted oscillations. Parasitic feedback can be dramatically reduced by careful layout and signal flow.

The simplest way to illustrate the undesirable impact of feedback is to consider a common emitter amplifier design shown in Fig. 8.2. It can be seen that poor layout can result in significant collector-to-base feedback, degrading gain and changing the input and output impedance of the amplifier. This can be avoided by both careful layout and routing of the lines that carry the main RF signal. Feedback can be reduced significantly by placing the collector and base lines and their associated passive components a safe distance apart and by avoiding crossover of the lines.

A major mechanism of feedback in such an amplifier is coupling of the RF signal from the output to the input through the substrate. When using on-chip components for matching purposes, it is critical to avoid large capacitors or inductors to reduce coupling. Use of the appropriate type of passive components is also very im-

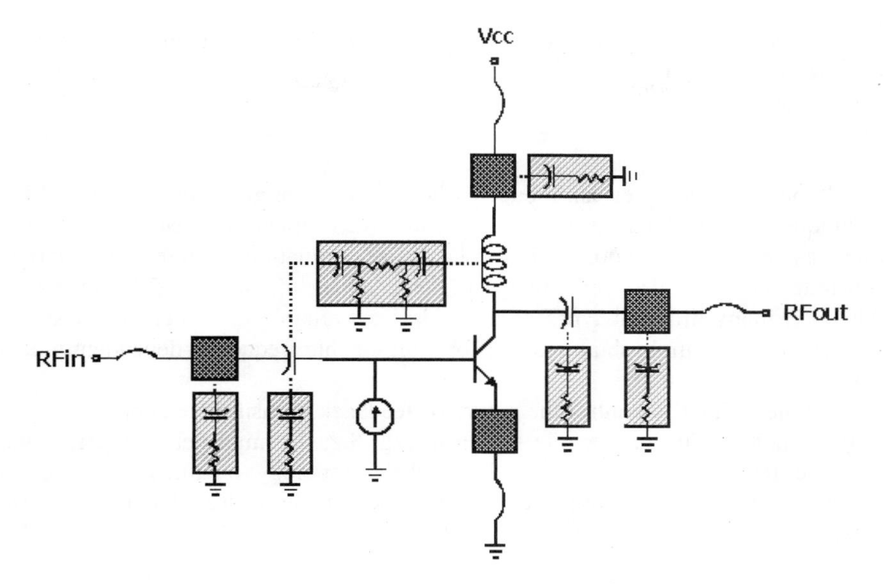


**Fig. 8.2.** A simple common emitter amplifier with parasitics modeled in the form of resistor and capacitor networks.

portant. For example, metal–insulator–metal (MIM) capacitors have far less substrate parasitics than MOS capacitors and, thus, are much more suitable for matching purposes [1]. Additionally, inductors with thick metal implemented on some type of substrate-isolating surface can help reduce substrate coupling and loss [1, 2]. Coupling can be further reduced by routing the RF signal on the top metal layers and using the top metal layers for implementing bond pads. Figure 8.3 demonstrates the capacitive coupling from a bond pad to the substrate. As shown in Eqn. 8.2, the

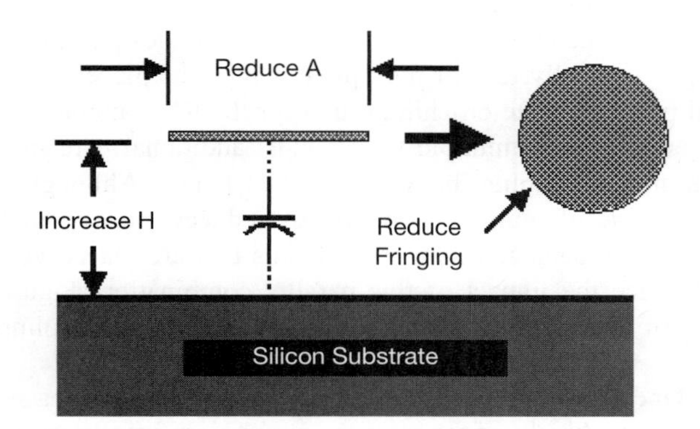


**Fig. 8.3.** Bond pad model and the capacitive components.

designer can reduce capacitance coupling to the substrate by reducing the area and increasing the height.

$$C = eA/h \tag{8.2}$$

Although the designer can reduce the substrate coupling by using many of the techniques described earlier, one would still have to model the feedback as accurately as possible for a good design-to-hardware correlation. Use of an appropriate substrate model is also very important for adequate modeling of the feedback. Although many different types of methods are available to model the substrate, we have used a simple shunt resistor for a reasonable frequency-dependent model [3].

Designers can also control the nature of feedback by using the appropriate amplifier topology. In the example shown in Fig. 8.2, the amplifier is experiencing negative feedback [4]. If a cascade amplifier topology is utilized instead on the common emitter, the amplifier will be experiencing positive feedback instead [5].

### 8.2.2. Grounding

Grounding is critical in successful design of any RF and microwave components. The ground reference is an inherent requirement for performing simulations and the proper allocation and modeling of a ground path can greatly impact the accuracy of the design. Poor on-chip grounding can have deleterious effects on the fabricated IC performance, leading to deviations in design-to-hardware correlation. The designer must identify the location of the RF ground and carefully model any parasitic inductances that may exist in the ground path. This location is typically the ground plane of the reference board. This will translate to a nonideal ground on the IC because the path from the board ground to the ground on the chip includes the inductance and resistance of the board, package paddle, or pin, and the down-bond connecting the chip ground to the paddle or pin. Issues regarding the package and its type will be further discussed in the next section.

The AC ground becomes less significant when the design migrates to a differential topology. Typically, use of multiple down-bonds placed in parallel can ensure availability of adequate on-chip ground for the RF components of the receiver. However, when using multiple wire bonds, additional care must be taken in modeling the mutual coupling between adjacent bonds. Although it is correct to assume that the use of multiple down-bonds reduces the overall ground inductance, the mutual inductance between the bonds that are placed very close to one another can reduce the impact of this parallel combination. Figure 8.4 shows a simple model of a bond wire and the outcome of mutual coupling between the two.

Low inductance grounding can be very helpful in reducing cross-talk. The isolation between various blocks of the receiver can be significantly improved by use of

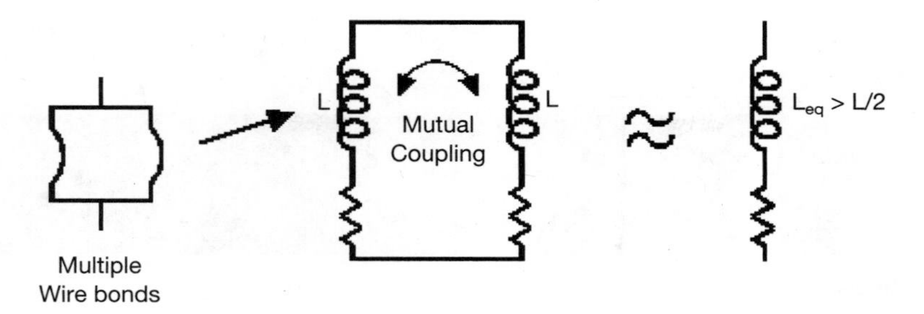


**Fig. 8.4.** Impact of mutual coupling on adjacent wire bonds.

low-inductance ground paths. Figure 8.5 shows a few examples of use and modeling of on-chip RF grounds.

### 8.2.3. Isolation

As mentioned earlier, physical distance between various receiver components decreases as the level of integration in an IC increases. Increasing the density of interconnection, active, and passive components inside an IC reduces the isolation between critical signals. Figure 8.6 illustrates the workings of deep trenches and isolation rings on improving isolation between two transistors.

Use of deep-trench isolation and isolation rings can help to improve poor isolation but this may not be enough. Some designers rely on careful frequency planning and unique architectures to reduce the impact of poor isolation on overall receiver performance.

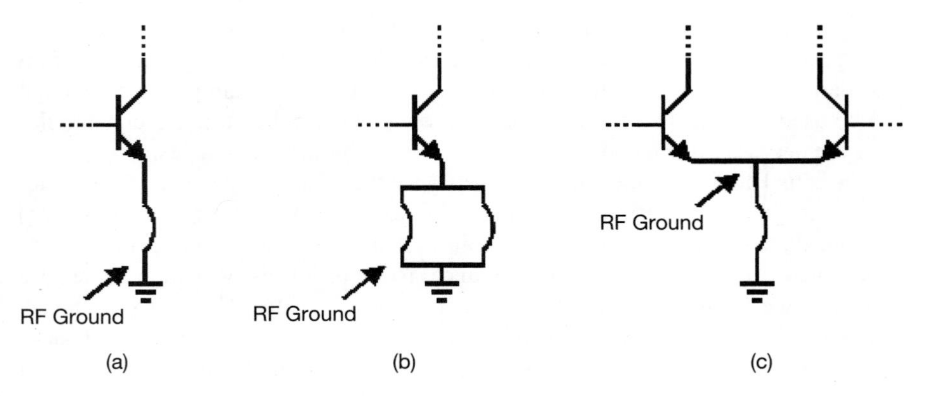


**Fig. 8.5.** Various RF grounds. (a) High-inductance RF ground for a single-end amplifier located at the far side of the wire bond. (b) Low-inductance RF ground using two wire bonds. (c) RF ground of a differential amplifier unaffected by the wire bond inductance.

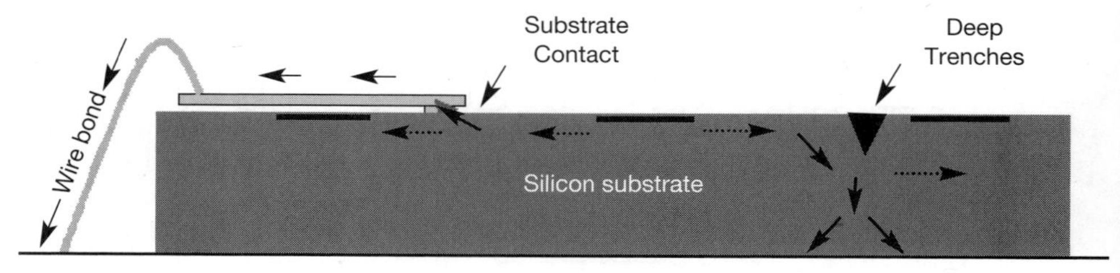


**Fig. 8.6.** Demonstration of RF surface currents and the impact of deep trenches and grounded substrate contacts on the current flow and direction.

## 8.3. PACKAGING CONSIDERATIONS

Packaging is very important in realizing an end-product that is feasible and can easily be integrated into a system. With the migration of applications into higher frequencies, the RF and microwave performance of packages are becoming more and more important and essential to the success of the design process. There are various packages available that can be considered for receiver implementation. The leaded plastic packages are the earlier generation of packages that were used for most receivers. However, with advances in packaging technology there are now many other type of packages available. A major problem with leaded packages is the inductance associated with the lead. This is eliminated by the use of leadless packages that reduce this inductance significantly. Another characteristic of a good package for RF applications is the paddle. To accommodate a good on-chip RF ground, the ground must be connected through down-bonds to the package paddle, which is soldered onto the board assembly ground. Without a paddle, the ground connection must be routed through longer bond wires and package pins, increasing the ground inductance.

Figure 8.7 shows an example schematic used to model the RF ground path from the board to the circuit implemented on the chip. This is a very simplistic model that can be used with relative accuracy up to a few gigahertz. Electromagnetic simulations are necessary for modeling such transitions at higher frequencies.

The actual amount of inductance or resistance that is used is dependent on the actual dimensions of the various components. For example, a 1 mil wire bond has a typical inductance of 0.9 nH/mm, but this value changes with the diameter of the wire bond. There are also some minor capacitances to the paddle associated with the wire bond or down-bond that we have ignored here in our estimates. The package pin parasitics are very much dependent on the shape and length of the pin. If a leaded package is used, the inductance and resistance values used for modeling the package pin will be significantly increased. For the case of a package with a paddle, the parasitics are much less and depend on the dimensions of the paddle itself. Typically, a few tens of picohenrys is a good estimate for package paddle parasitics. Modeling of the board parasitics is very much dependent on the board layout. It is desirable that the board

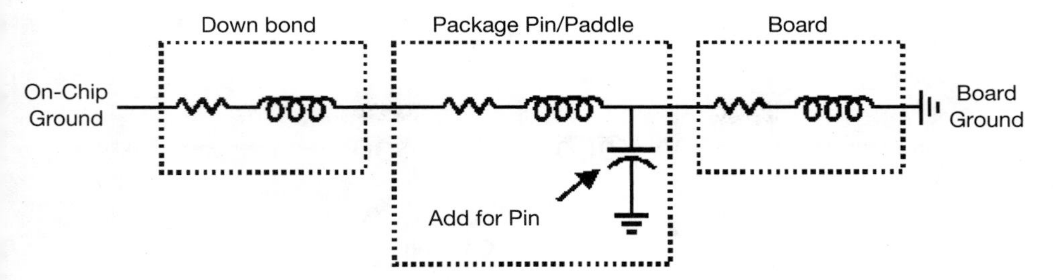


**Fig. 8.7.** A lumped-element model of the RF ground path from the circuitry on an IC to the board ground.

ground plane cover a large area and be well connected throughout different layers of the board with numerous grounds through vias. This would ensure a relatively constant ground potential on the board and reduce any parasitic effects. Figure 8.8 shows the different types of common packages used for RF ICs and modules.

### 8.3.1. Package Modeling

Package models are typically provided by the packaging houses and work pretty well up to a couple of gigahertz. For use above such frequencies, additional effort must be made to verify the quality of the models that are being used. Use of EM simulations can do this in an effective manner. One of the critical issues associated with packages is the pin-to-pin capacitance, which can directly impact the quality of isolation between adjacent pins. Another major parasitic component is the capacitance between the pins and the ground paddle or the ground of the board. This can impact RF matching or increase loss through the pin at high frequencies.

Resistive and inductive effects are also important in modeling pins in a package. When using leadless packages, the lead inductance is typically very small but it

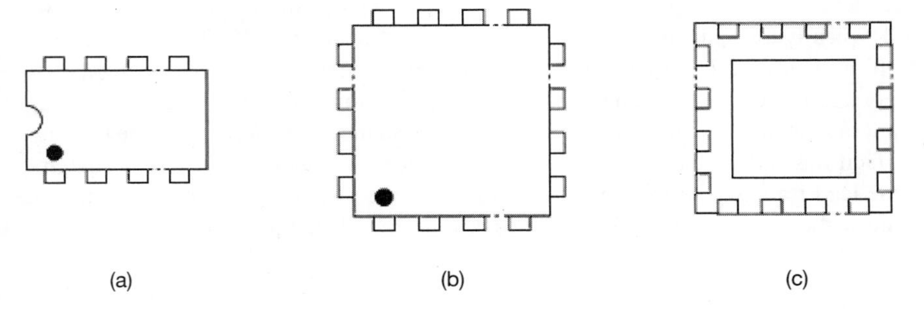


**Fig. 8.8.** Different package types. (a) Leaded shrink small outline package (SSOP) and thin small outline package (TSOP). (b) Leaded quad flat package (QFP). (c) Low-profile leadless package.

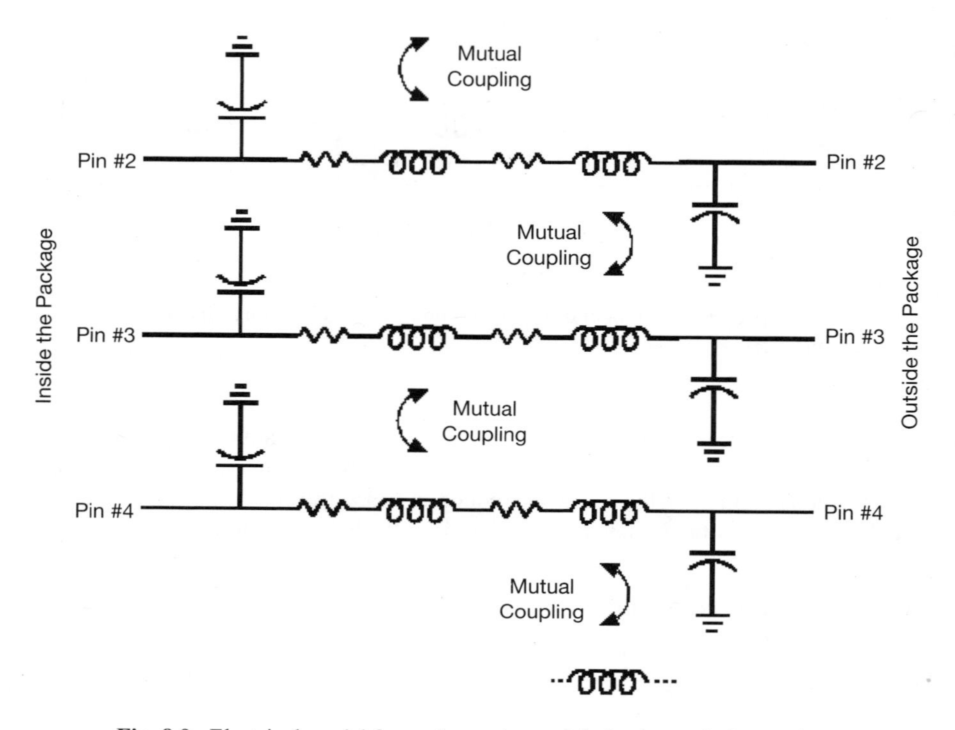


**Fig. 8.9.** Electrical model for package pins and their pin-to-pin interaction.

should still be considered during the simulations. Figure 8.9 shows an example of a package model.

### 8.3.2. Bonding Limitation

There are many design rules that are supplied by the packaging houses that must be considered when designing integrated receivers. As mentioned earlier, integration will typically translate into higher interconnection density in an IC. This can manifest itself in the use of larger packages that can accommodate more pins, leaving a relatively large distance between the pins and the die sitting inside the package. This can result in stretching the length of wire bonds beyond the limits defined by the packaging facilities and increasing the inductive parasitics of the wire bonds. Use of custom-made lead frames with multiple rows of pins is one of the ways to increase the pin count without increasing the size of the package and the length of the wire bonds.

## 8.4. CONCLUSION

In this chapter, we have introduced the reader to the various practical concepts required for the development of a monolithic, integrated multifunctional RF front-end. This chapter has addressed issues regarding physical implementation, parasitic modeling, and the discrepancies between design and hardware performance. All of these details are critical in the development of integrated receivers and receiver subsystems, starting from a design concept to a packaged IC. As the technologies progress towards higher-frequency applications, newer techniques might be required, but this chapter has laid the foundation for realization of higher-frequency receivers in silicon-based processes for commercial applications up to a few gigahertz.

## REFERENCES


1. S. Jenei, S. Decoutere, S. Van Huylenbroeck, G. Vanhorebeek, and B. Nauwelaers, "High *Q* inductors and capacitors on Si substrate," in *IEEE Silicon Monolithic Integrated Circuits in RF Systems,* pp. 64–70, 2001.
2. J. N. Burghartz, M. Soyuer, K. A. Jenkins, M. Kies, P. Dolan, K. Stein, J. Malinowski, and D. L. Harame, "RF components implemented in an analog SiGe bipolar technology," in *Bipolar/BiCMOS Circuits and Technology Meeting,* pp. 138–141, Sept. 1996.
3. J. Colvin, S. S. Bhatia, and K. K. O, "Effects of substrate resistances on LNA performance and bondpad structure for reducing the effects in a silicon bipolar technology," *IEEE Journal of Solid State Circuits, 34,* 9, Sept. 1999, pp. 1339–1344.
4. B. Banerjee, B. Matinpour, and J. Laskar, "Development of IEEE 802.11a WLAN LNA in silicon-based processes," in *MTT Symposium,* June 2003.
5. Q. Liang, G. Niu, J. D. Cresseler, S. Taylor, D. L. Harame, "Geometry and bias current optimization for SiGe HBT cascode low-noise amplifier," in *IEEE Radio Frequency Integrated Circuits Symposium,* pp. 407–410, 2002.

# 9

# FUTURE TRENDS

## INTRODUCTION

The previous chapters introduced a wide variety of advanced RF front-end circuits and systems, with an emphasis on high-frequency implementations in silicon-based process technologies. In this chapter, we introduce the readers to some futuristic applications in the area of wireless communications. With the evolution of technology, it would be interesting to see whether some, few, many, or any of these approaches become a reality. The rapid growth of wireless technologies coupled with the tremendous growth of silicon-based process technologies has opened up a vista of innovative applications. Whereas some of these can be targeted towards existing wireless applications, leading to more options and flexibility, others might offer a huge paradigm shift from the original line of thought related to existing wireless communication technologies. As the readers may find, some of these might be obsolete in the near future, whereas some of them might be viable a long time from now. It has always been a researcher's dream to make impossible things happen and extend the domains of fundamental knowledge and understanding. All of the following goals look challenging but they may possibly be tractable. As time progresses, we are slowly reaching a level at which it may be possible to observe a mixture of microwave circuit design concepts with electromagnetic phenomena, traditional analog circuit design principles, and digital logic, all in the same environment. All of these would implement the entire world of diverse design principles in a tiny piece of silicon.

## 9.1. CMOS CELLPHONES

There has been a great deal of debate regarding the feasibility of CMOS technology for cellular RF applications. This potential low cost candidate does not come with-

*Modern Receiver Front-Ends.* By J. Laskar, B. Matinpour, and S. Chakraborty
ISBN 0-471-22591-6 © 2004 John Wiley & Sons, Inc.

out costs. With the continued scaling of CMOS devices to the deep-submicron regime, different issues come into consideration, such as: a) yield for production, b) degradation due to hot electron effects and associated reliability issues, c) lossy substrate and low quality factor of passives.

It is unlikely that we will see a fully integrated RF-CMOS radio for mainstream cellular telephony. The primary limitations include the required efficiency and linearity for the transmitter coupled with the required isolation and switching requirements for the receiver. That said, we see increased migration of RF-CMOS through the IF stage and through some of the key superheterodyne block functions. One can postulate that GaAs will find a niche in front-end PA and switching, facing increasing competition from SiGe. However, SiGe will face increased competition from CMOS for a majority of IF circuit blocks.

## 9.2. MULTIBAND, MULTIMODE WIRELESS SOLUTIONS

More options are usually considered to be better. The next step in the development of wireless solutions is a truly multiband, multimode solution. This will enable people to effectively utilize the bandwidth and tune to an appropriate data rate whenever needed, with freedom of movement between countries.

However, without much digression, let us focus on our main stream of thought as presented earlier in this book. A fully multiband system requires much careful system planning and architecture analysis. Direct conversion architecture is a potential solution for a fully monolithic implementation, as it eliminates complicated IF planning. For a multiband direct conversion architecture, one could wisely group applications that utilize a "DC-free" modulation scheme. However, if this requirement is not satisfied, one has to find a trade-off between the filter cutoff frequency and signal-to-noise ratio (SNR) for specific applications. Figure 9.1 illustrates a potential system architecture for a multiband solution. It is important to determine the position of the front-end switch. LNAs for these applications could be designed in a concurrent fashion [1], which can be applicable for the frequency bands under consideration. One can also think about designing separate LNA hardware for different frequency bands under consideration, as shown in Fig 9.1. In both cases, the semiconductor area is about the same, as both of these options would have to use a similar number of on-chip inductors. Migration of the entire radio, excluding the front-end components such as PA, LNA, and switches, at frequencies less than 10 GHz, is a near-term reality for most of the existing wireless technologies in silicon-based processes.

Another possibility in system integration is to transfer all the multiband options to the front-end passive components and design a multiband filter and antenna. This can be developed in ceramic substrates that provide high quality factor ($Q$) passives. However, this implementation is also quite challenging with respect to sufficient rejection when the frequency bands are very close (for example, the 5.2 GHz and 5.8 GHz bands in the IEEE 802.11a standard).

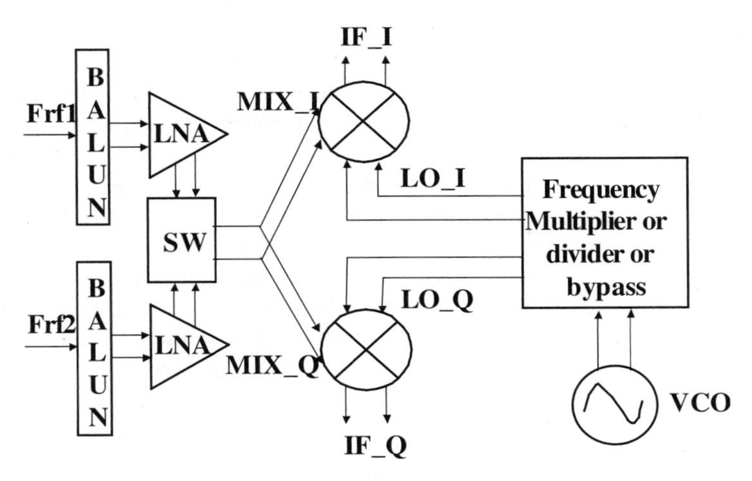


**Fig. 9.1.** System architecture for a potential multiband application.

Considering this, the curious reader might ask, "when will it be possible to have a fully CMOS multiband cellular phone?" In reality, only time and future technological development will tell. Many industries and research organizations are trying to discover a solution for these applications.

## 9.3. 60 GHZ SUBSYSTEMS IN SILICON!

Let us take our mind off the existing wireless communication frequency bands and details of front-end architecture for a moment, and think about RFIC designs at a frequency of 60 GHz. Recent reports of 200 GHz cut-off frequency transistors in silicon germanium BiCMOS technology [2, 3] have opened up an interesting set of possibilities for very high frequency RFICs in silicon-based technologies. This shows the tremendous potential of silicon-based technologies (silicon germanium) as a possible challenge to the existing pHEMT-based implementations [4, 5]. We would like to remind the interested reader about our earlier discussions of the excellent high-frequency mixing performance of majority-carrier-based Schottky barrier diodes. A standalone Schottky barrier diode-based mixer is capable of operating at a much higher frequency than the cutoff frequency of the active devices in the technology due to its reduced parasitics. However, the mixer needs to be integrated with the low-noise amplifier in the RF front-end, and for 60 GHz operation of the LNA, the desired cut-off frequency of the active devices is around 200 GHz (following the rule of thumb, $F_T/3$, for analog circuit design).

Some of the approaches presented earlier in this book, especially subharmonic mixing and direct conversion, are quite relevant for the development of a receiver at a very high frequency. Apart from the challenges in designs, simulations, and layout optimization, there exists an even tougher domain of characterization of the developed circuits and systems at 60 GHz. At such a high frequency, the wavelength becomes quite small and one can consider a fully functional RF front-end with a smaller total dimension of less than a wavelength. When ICs can be developed with such a high-performance, state-of-the-art process technology, passive components cannot be left far behind. One can realize on-chip filters for band-select filtering, with an antenna on the reverse side of the silicon wafer. This is the perfect example of classical techniques from traditional high-frequency electromagnetics mixing with high-frequency realization of analog circuits for a fully monolithic implementation for wireless solutions, all in a silicon-based technology. The low-frequency analog and digital circuits are also integrated on the same die to realize a fully monolithic system-on-chip solution.

## 9.4. INTERCHIP COMMUNICATIONS

Having introduced the 60 GHz silicon-based wireless receiver as a potential futuristic application, let us concentrate more on the paradigm shift toward more microlevel communication applications. Along with the tremendous development in the wireless communications applications, a different school of thought has evolved for applying similar techniques in a much smaller environment for wireless chip-to-chip communication. Interchip communication becomes a real challenge as the technology CMOS scaling reaches gate lengths below 0.1 μm, at which the delay of the local interconnect becomes a relatively significant portion of the total delay. For a 1 mm interconnect length, the wire delay is greater than the transistor delay at 0.1 μm gate length itself [6]. Thus, for the global interconnects, the delay is already a concern for current-generation designs. For the local interconnects, the wiring delay would become larger than the gate delay at a 0.065 μm gate length. This motivates one to investigate solutions in which one can perceive chip-to-chip communication in a wireless fashion. The idea can only succeed if it can be accomplished with minimal overhead in the existing system, in terms of power consumption and circuit complexity. Due to the proliferation of the Bluetooth standard for short-range wireless communications in the recent past, it is quite tempting to refer to it as "microtooth"—a microlevel realization of Bluetooth.

There have been reports of wireless interconnect systems that transmit and receive RF signals across a chip using integrated antennas, receivers, and transmitters [7]. This approach is based on a transmitter integrated with on-chip antennas, all operating at a frequency of 15 GHz in a 0.18 μm CMOS process. Figure 9.2(a) shows an illustration of this wireless interconnect system; the open loop transmitter and the clock receiver are shown in Fig. 9.2(b). This approach is very similar to the far-field broadcasting techniques used in early days of wireless communications. However, the efficient transmission and reception of RF/microwave signals in free space require the size of the antenna to be compatible with their wave-

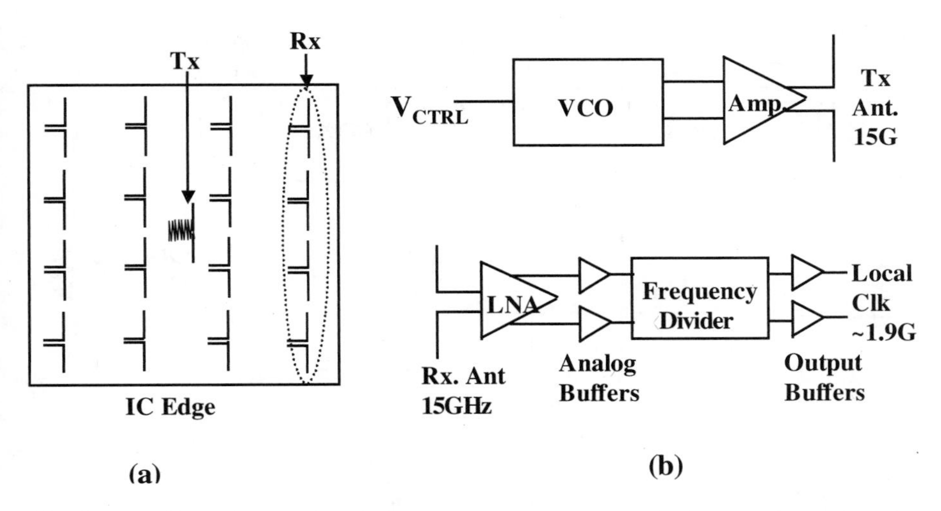


**Fig. 9.2.** An illustration of (a) a wireless interconnect system, and (b) transmitter and receiver system blocks for a wireless interconnect system.

lengths. As the CMOS device dimensions continue to scale down, the cutoff frequency is expected to reach 100 GHz in the near future. However, even at this frequency, the optimal size of the antenna is on the order of 1 mm$^2$. This dimension is too large to be comfortably implemented in future ultralarge-scale integration (ULSI).

Microwave transmission in guided media, such as the microstrip transmission line (MTL) or coplanar waveguide (CPW), is known to exhibit low attenuation up to 200 GHz. A comparative study between CPW and conventional ULSI interconnects has been reported in [8]. It has been observed that signals transmitted through a 1-cm-long CPW experience a low loss of 1.2 dB at 100 GHz, across 50–150 GHz range, and a dispersion of less than 2 dB, whereas the loss of conventional ULSI interconnects is about 60 dB/cm and 115 dB/cm for 1 $\mu$m and 0.1 $\mu$m, respectively, with a frequency dispersion of 30–40 dB across the frequency range. These low loss and low dispersion characteristics of the guided media have motivated the development of an RF/wireless interconnect system in which the output signals from several ICs can be up-linked to MTL or CPW lines via transmission-capacitive couplers ($C_T$), and then down-linked via reception capacitors ($C_R$) to input ports to fulfill the interconnect function. Since the communication distance is relatively short (several centimeters apart), the sizeable "far-field" antenna can be substituted for much smaller "near-field" capacitive couplers. Although the size is somewhat large, the CPW or MTL can be used as an "off-chip" but "in-package" transmission medium and shared by multiple ULSI I/Os.

Based on the above considerations, the proposed RF/wireless interconnect system can be depicted as shown in Fig. 9.3, in the form of a miniature WLAN. Modern code division (CDMA) and/or frequency division multiple-access

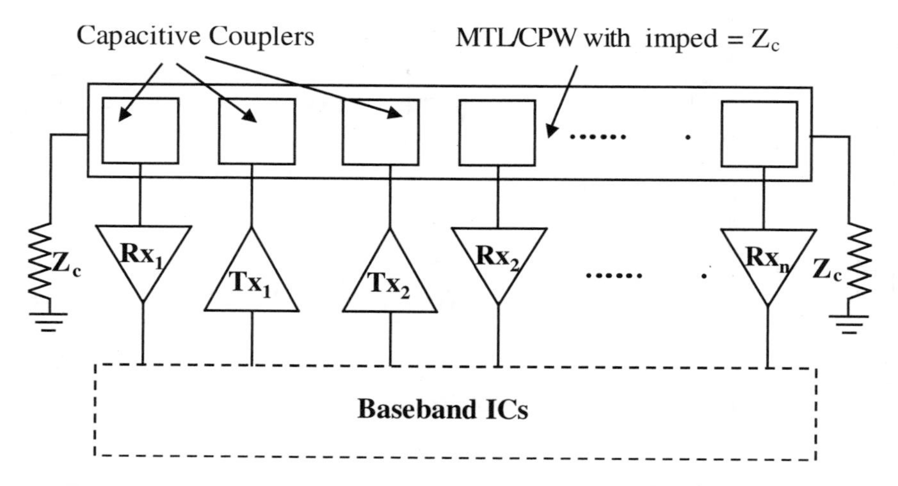


**Fig. 9.3.** A representative RF/wireless interconnect channel with multiple I/Os.

(FDMA) algorithms can be used effectively to mitigate the effects of the undesired cross-channel interface within the shared medium. With orthogonal-coded and/or frequency-filtered RF transceivers, a passive MTL or CPW can be very suitable for relaying ultrabroadband signals up to 150 GHz [8]. Figure 9.4 is an illustration of physical representation of the above RF/wireless interconnect scheme. Figure 9.5 shows a representative RF interconnect channel shared by multiple I/Os ($Rx_i$ to $Rx_j$, and $Tx_i$ to $Tx_j$). The channel consists of a uniform and homogeneous CPW with its $Z_c$ equal to 50 Ω. The signal line of the CPW is terminated with $Z_c$ to avoid signal reflection.

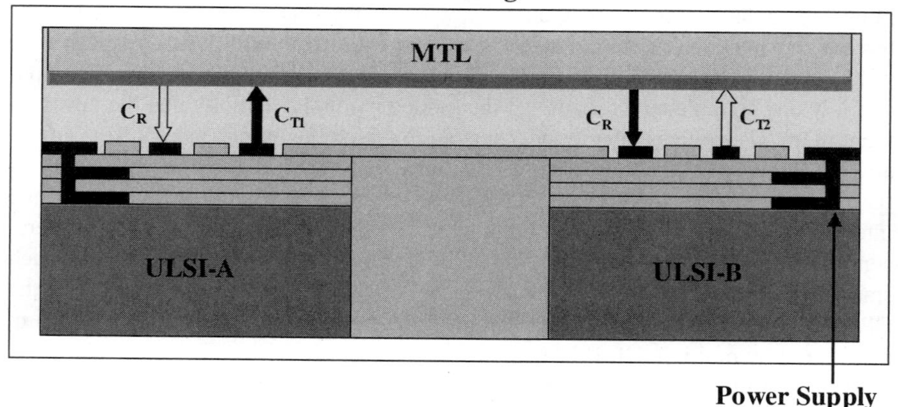


**Fig. 9.4.** Physical representation of the RF/wireless interconnect scheme.

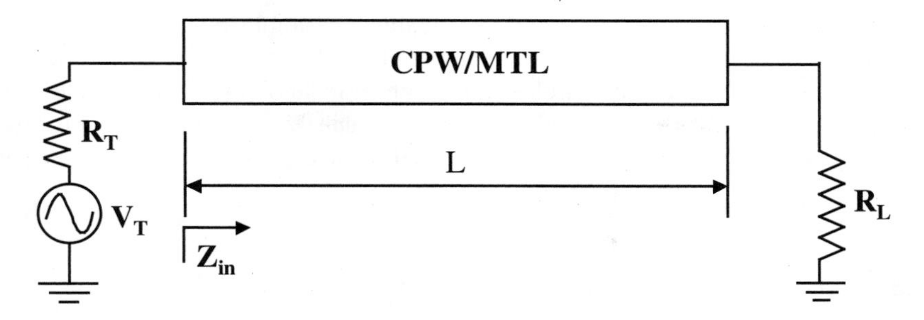


**Fig. 9.5.** Termination of the CPW/MTL to minimize frequency dispersion effects.

Figure 9.6 shows the equivalent circuit for a representative transmitter–receiver loop. The receiving signal strength can be correlated to the transmitted signal strength by

$$V_{\text{rec}} = \frac{\dfrac{Zc}{2} R_{\text{R}}}{\left(\dfrac{Zc}{2} + R_{\text{T}} - j\dfrac{1}{\omega C_{\text{T}}}\right)\left(\dfrac{Zc}{2} + R_{\text{R}} - j\dfrac{1}{\omega C_{\text{R}}}\right)} V_{\text{T}} e^{-j\beta d} \tag{9.1}$$

where
$V_{\text{T}}$ = Transmitter signal voltage
$d$ = distance between $Tx_{\text{i}}$ and $Rx_{\text{j}}$
$R_{\text{T}}$ = $Tx_{\text{i}}$'s output resistance
$R_{\text{R}}$ = $Rx_{\text{j}}$'s input resistance

As is evident from Eqn. 9.1, one should choose $R_{\text{T}}$ and $R_{\text{R}}$ to be much greater than $Z_{\text{c}}$ to preserve CPW's characteristic impedance and choose $R_{\text{R}} \gg j/\omega C_{\text{R}}$ and $R_{\text{T}} \gg j/\omega C_{\text{T}}$ to obtain a dispersion-free $V_{\text{rec}}$. Larger $C_{\text{T}}$ and $C_{\text{R}}$ are preferable for better

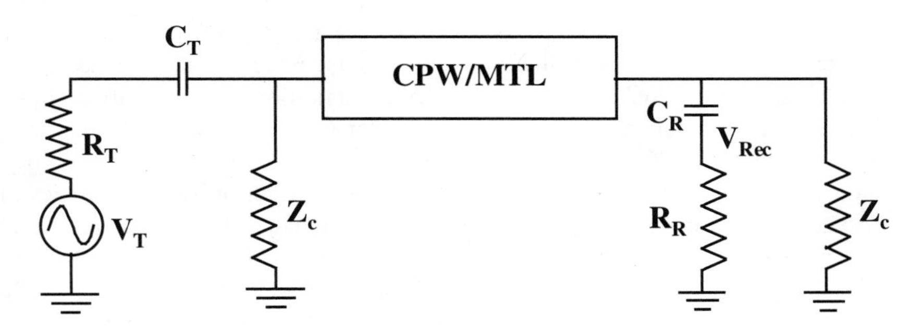


**Fig. 9.6.** Equivalent circuit for transmitter–receiver loop in the wireless interconnect system.

signal coupling between transceivers and the transmission line. The sizes of $C_T$ and $C_R$ must be made small enough to be compatible with the future ULSI fabrication. However, the success of this wireless interconnect technology will only be evident when the RF part of the transceiver subsystem can be implemented with much smaller power and area requirements compared to the digital counterpart of the subsystem.

## 9.5. ULTRAWIDEBAND COMMUNICATION TECHNOLOGY

Ultrawideband (UWB) radio systems have attracted a lot of attention during the last few years. UWB radio technology differs from conventional narrowband radio and spread-spectrum technologies in that the half power bandwidth of the signal is typically from 25% to 100% of the center frequency. Instead of transmitting a continuous carrier wave modulated with information (or with information combined with a spread code), which determines the bandwidth of the signal, a UWB radio transmits a series of very narrow impulses. The UWB radio has a distinct advantage over narrowband radio systems in resistance to signal degradation by multipath propagation. In environments having reflecting objects, the continuous carriers of narrowband systems are susceptible to destructive interference due to multipath signals. In an UWB radio system, the multipath signal energy arrives at a different time than the direct path energy, and cannot create destructive interference. A multiple correlator system can recover the multipath energy to enhance the system performance [9].

These systems use very low transmission power, spread over a bandwidth of several gigahertz. The very low transmission power and the large bandwidth usage enable UWB radio systems to coexist with other narrowband systems over the same frequency band without interfering with the narrowband systems. However, the narrowband systems may cause interference that can jam the UWB receiver. There have been efforts to develop narrowband interference suppression techniques for UWB radio systems [10].

Hardware developments for UWB technology have been also reported in the open literature [10]. Figure 9.7 shows a block diagram for the UWB receiver under consideration. The function of the correlator module is to recover the energy in the transmitted impulses and present a representative signal to the baseband module. The baseband module performs the functions of pulse integration and handling the timing offset during synchronization and reconstruction of the information signal. The timing module generates the precision master clock for all signal processing operations and provides the facility for the baseband module to time-shift the trigger signal to the correlator for synchronization. After synchronization is achieved, the pseudorandom code is included in the timing signal to the correlator, which strips off the transmitted signal's pseudorandom coding. Till this point, the correlator output signal contains the information modulation and a DC component. The timer–correlator combination provides the critical element for reception of the time-modulated ultrawideband signal.

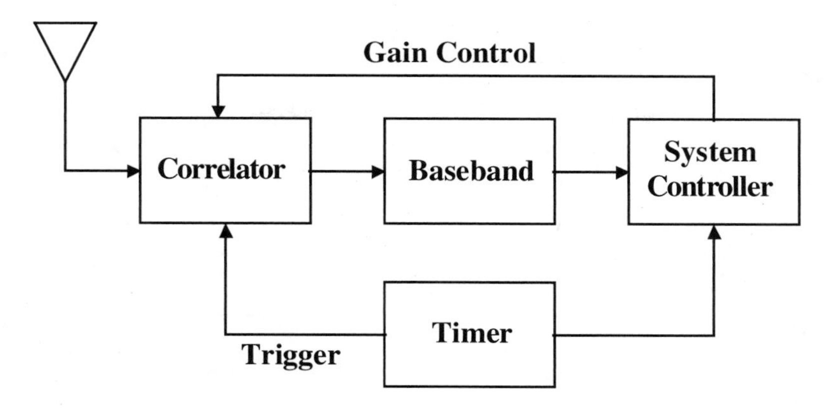


**Fig. 9.7.** Ultrawideband receiver block diagram.

## 9.6. DIVERSITY TECHNIQUES

Along with the needs for high data rate, low power, low cost, and portability, the next-generation wireless communication standards are required to operate reliably in different environments, such as macro-, micro-, and picocellular, as well as urban, suburban, and rural, both indoors and outdoors. Thus, they need to be bandwidth-efficient and capable of being deployed in diverse environments. A big deterrent to such systems is multipath fading, which could result in up to 10 dB improvement in SNR, for a BER reduction from 1e-2 to 1e-3. One of the efficient methods to combat multipath fading is transmitter power control, which requires a high transmitter dynamic range. Other techniques to improve diversity include time and frequency diversity and time interleaving, along with error correction coding and spread spectrum techniques. Time interleaving results in large delays when the channel is slowly varying, and spread spectrum becomes less effective when there is relatively small delay spread in the channel.

Antenna diversity is a very widely used and effective technique to combat effects of multipath fading in a scattering channel. Classical approaches use multiple antennas in the receiver side and perform combination or selection of the better path to either improve the quality of the signal or select the signal with better quality. Problems associated with receiver diversity thus include the cost, size and power of the receiver units. Thus, it is quite beneficial to use diversity techniques at the base stations. Methods incorporating a delay diversity scheme suggest the transmission of copies of the same symbol through multiple antennas at different times, and use of equalizers to resolve multipath distortion and obtain diversity gain. Another approach is to use space–time trellis coding, in which symbols are encoded according to the antennas through which they are simultaneously transmitted, and are decoded using a maximum likelihood decoder.

Within the scope of this book, we present a classical two-branch "maximal-ratio receive combining" (MRRC) scheme in Fig. 9.8. It should be noted that at a given

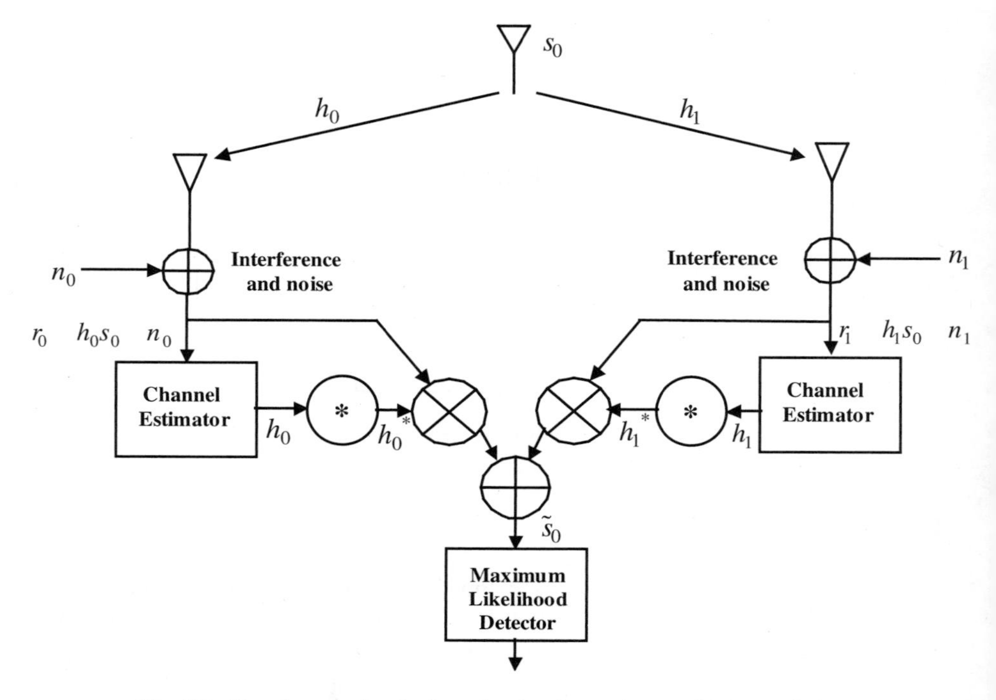


**Fig. 9.8.** Two-branch classical maximal-ratio receive combining technique.

time, signal $S_0$ is sent from the transmitter. The channels between the transmit antennas and the receive antenna are denoted by $h_0$ and $h_1$, where

$$h_0 = \alpha_0 e^{j\theta_0}, \qquad h_1 = \alpha_1 e^{j\theta_1} \tag{9.2}$$

With noise and interference added at the two receivers, the resulting received signals can be represented by

$$r_0 = h_0 s_0 + n_0, \qquad r_1 = h_1 s_0 + n_1 \tag{9.3}$$

Assuming Gaussian distribution for the noise, a maximum likelihood decision rule can be applied at the receiver, and the appropriate symbol can be chosen depending on the Eucledian distances from each of the received symbols. The mathematical formulation to select symbol $s_i$ could be represented as

$$d^2(r_0, h_0 s_i) + d^2(r_1, h_1 s_i) \le d^2(r_0, h_0 s_k) + d^2(r_1, h_1 s_k), \ \forall\ i \neq k \tag{9.4}$$

A modification of the diversity scheme shown in Fig. 9.8 can be represented as shown in Fig. 9.9 [11]. The scheme shown in Fig. 9.9 uses two transmit antennas and one receive antenna, and performs three fundamental functionalities: (1) en-

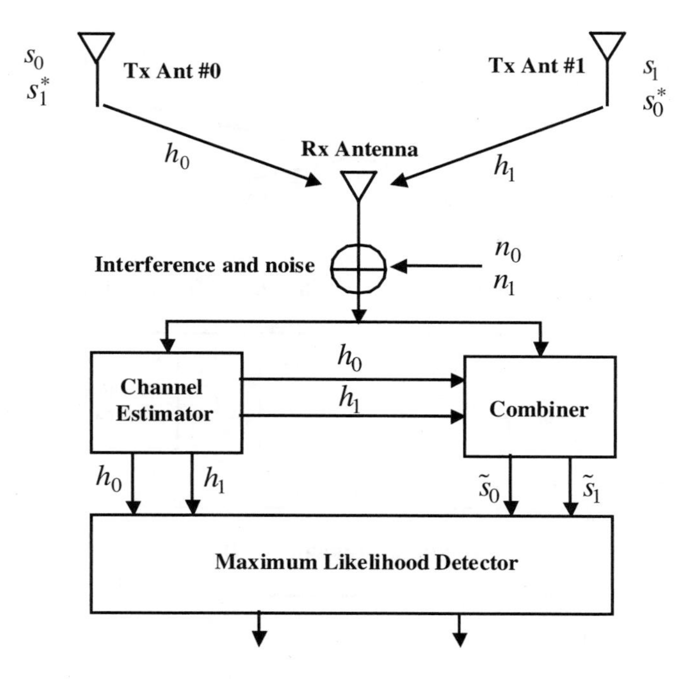


**Fig. 9.9.** Two-branch transmit diversity with one receiver.

coding and transmission sequence of information symbols at the transmitter, (2) combining scheme at the receiver, and (3) decision rule for maximum likelihood receiver. For a particular symbol period, two signals ($s_0$ and $s_1$ respectively) are simultaneously transmitted from the two antennas. In the next symbol period, the complex conjugates of these symbols are sent (i.e., $-s_1^*$, and $s_0^*$, respectively). Hence, the coding can be thought of as space–time coding. It could be performed by space–frequency coding by using two adjacent carriers instead of adjacent time slots.

The scheme shown in Fig. 9.9 can easily be extended to cover multiple receivers. One such scenario is presented in Fig. 9.10.

## 9.7. CONCLUSION

In this chapter, we have presented some of the potential future applications and development of high-frequency transceiver subsystems. One school of thought is to continuously improve the integration level and novel features in existing transceiver subsystems, while the other is to focus and explore research and visionary ideas and applications for wireless technologies. As the technology is rapidly progressing, new wireless technologies can also bring new application possibilities. It would be quite interesting to see how many of these actually materialize.

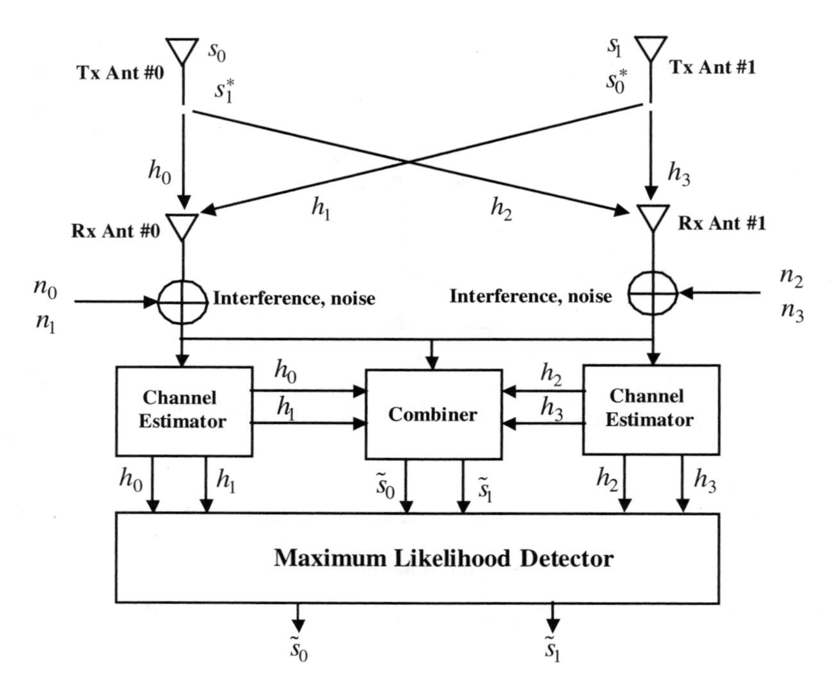


**Fig. 9.10.** Two-branch transmit diversity with two receivers.

## REFERENCES


1. H. Hashemi and A. Hajimiri, "Concurrent dual-band CMOS low noise amplifiers and receiver architectures," in *IEEE VLSI Circuit Digest of Technical papers,* 2001, pp. 247–250.
2. *http://www.ibm.com/news/2001/06/25.phtml.*
3. D. Greenberg, B. Jagannathan, S. Sweeney, G. Freeman, and D. Ahlgren, "Noise performance of a low base resistance 200 GHz SiGe technology," in *IEEE Electron Devices Meeting (IEDM) Technical Digest,* 2002, pp. 22.1.1.–22.1.4.
4. D. Streit, R. Lai, A. Oki, and A. G. Aitken, "InP HEMT and HBT applications beyond 200GHz," in *14th Indium Phosphide and Related Materials Conference,* (IPRM), 2002 pp. 11–14.
5. K. L. Tan et al., "60GHz Pseudomorphic $Al_{0.25}Ga_{0.75}As/In_{0.28}Ga_{0.72}As$ Low-noise HEMT's," *IEEE Electron Device Letters, 12,* 1, 23–25. January 1991.
6. J. D. Meindl et al., "Interconnecting device opportunities for gigascale integration (GSI)," in *IEEE Electron Devices Meeting (IEDM) Technical Digest,* 2001, pp. 23.1.1–23.1.4.
7. B. A. Floyd, C.-M. Hung, and Kenneth, K. O. "Intra-chip wireless interconnect for clock distribution implemented with integrated antennas, receivers, and transmitters," *IEEE Journal of Solid State Circuits, 37,* 5, 543–552, 2002.

8. M. F. Chang, V. P. Roychowdhury, L. Zhang, H. Shin, and Y. Qian, "RF/wireless interconnect for inter- and intra-chip communications," *Proceedings of the IEEE, 89,* 4, 456–466, 2001.
9. D. Dickson and P. Jett, "An application specific integrated circuit implementation of a multiple correlator for UWB radio applications," in *IEEE Military Communications Conference Proceedings, 1999,* Vol. 2, pp. 1207–1210.
10. I. Bergel, E. Fishler, and H. Messer, "Narrow-band interference suppression in time-hopping impulse-radio system," in *Proceedings of IEEE Conference on Ultrawideband Systems and Technologies, 2002,* Vol. 2, pp. 303–308.
11. S. M. Alamouti, "A simple transmit diversity technique for wireless communications," *IEEE Journal on Selected Areas of Communications, 16,* 8, 1451–1458, October 1998.

# INDEX



*Modern Receiver Front-Ends.* By J. Laskar, B. Matinpour, and S. Chakraborty
ISBN 0-471-22591-6